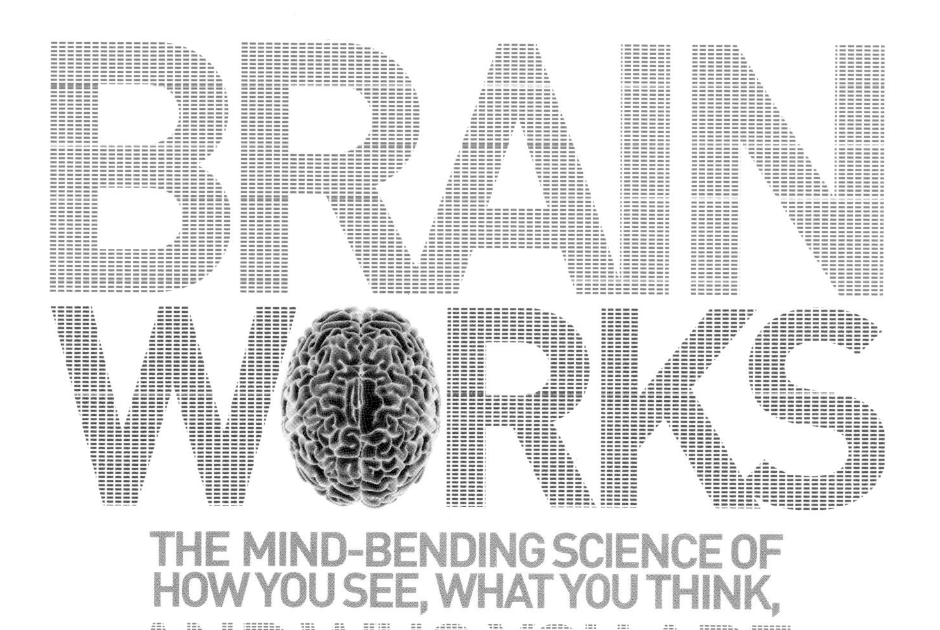

NATIONAL GEOGRAPHIC

WASHINGTON, D.C.

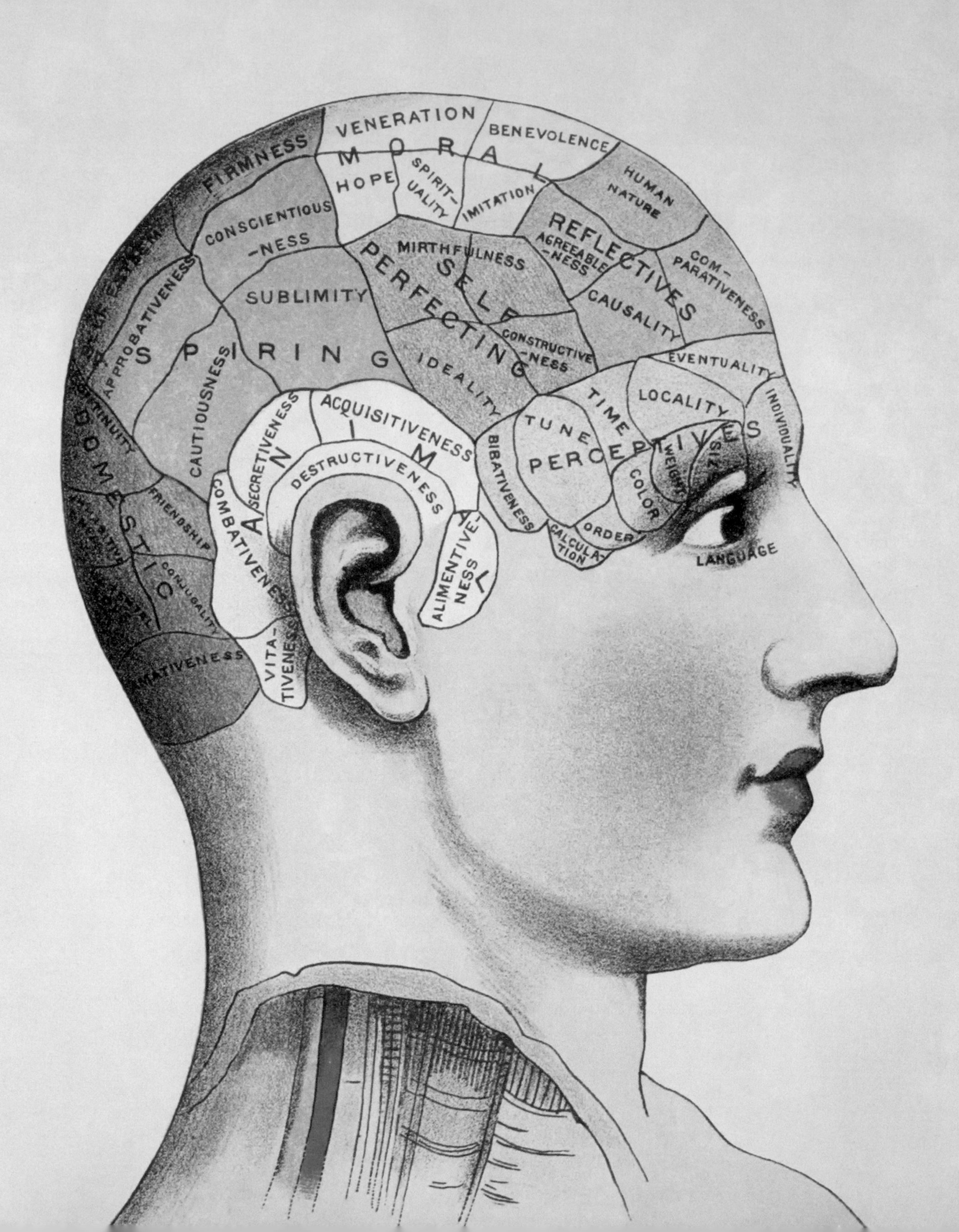

VENERATION
BENEVOLENCE
M O R A L
FIRMNESS
HOPE
SPIRIT-UALITY
IMITATION
HUMAN NATURE
CONSCIENTIOUS-NESS
REFLECTIVES
AGREEABLE-NESS
COM-PARATIVENESS
MIRTHFULNESS
SELF PERFECTING
SUBLIMITY
CAUSALITY
APPROBATIVENESS
S P I R I N G
CONSTRUCTIVE-NESS
EVENTUALITY
IDEALITY
CAUTIOUSNESS
LOCALITY
TIME
INDIVIDUALITY
ACQUISITIVENESS
A N I M A L
TUNE
PERCEPTIVES
SECRETIVENESS
DESTRUCTIVENESS
BIBATIVENESS
WEIGHT
SIZE
COLOR
CONTINUITY
DOMESTIC
FRIENDSHIP
COMBATIVENESS
ORDER
CALCULA-TION
LANGUAGE
ALIMENTIVE-NESS
CONJUGALITY
VITA-TIVENESS
AMATIVENESS

Contents

Pages 2-3: Zigzag disks appear to move, thanks to afterimages of complementary colors in our peripheral vision.

Facing page: Ingenuity, connubial love, kindness, intuition, hope, and many more mental faculties are pinpointed on a map of the brain used by the pseudoscience of phrenology in the mid-19th century.

BRAIN BENDER

I'm a professional illusionist. I fly without strings or camera tricks. I make it snow. I've made the Statue of Liberty and a Lear jet disappear. I've cut myself in half. And at every show I make 13 volunteers from the audience vanish. Audiences entrust me with their perceptions, knowing I'll bend them into interesting shapes and suspend their disbelief to provide, for 90 minutes, something that looks a lot like the miraculous. I create illusions that use storytelling, music, and psychology to evoke emotions and amazement in the viewer's mind. The effect of experiencing this kind of magic is not confusion but wonder. • Magicians have been called the scientists of show business. The stage is our laboratory, and through trial and error we've learned a lot about the mysterious inner workings of the brain. We've figured out that, with some skill and misdirection, we can get an audience to focus its attention in the right place and at the right time so that we can create the illusion of magic. In fact, these illusions are created not on the stage but in the brain. • Perception is influenced by many things, including cultural biases and beliefs, and a skilled magician takes advantage of these things to accomplish the appearance of the miraculous. An audience watching magic

DAVID COPPERFIELD

Behind the curtain

Zan Zig performs four vignettes, including decapitation and levitation, in an 1899 poster. Neural processing underlies a magician's tricks.

in a theater in New York or Paris or Los Angeles has a very different orientation about the performance than an audience from a culture that believes in magic as practiced by shamans and medicine men.

In a famous story, Jean-Eugène Robert-Houdin, a French magician generally considered the father of modern magic (and the inspiration for Ehrich Weiss to rename himself Houdini), was called on by France to quell a political uprising in Algeria, which France controlled. A sect of supposed holy men called Marabouts were using trickery to make the Algerians believe they had supernatural powers. Trading on the devotion their chicanery inspired, they were inciting their countrymen to revolt and cut their ties to France. The French government sent Robert-Houdin to Algeria with the directive to "outmagic" the Marabouts.

Robert-Houdin arrived in Algeria with a small metal box. He put the box on the ground and defied the strongest, largest Marabout to lift it. The Marabout who accepted the challenge, an enormously strong weight lifter type, grabbed the box and saw his confidence magically change to confusion, then embarrassment. He couldn't budge it. Yet Robert-Houdin, who was a slender guy, had been carrying the thing moments before.

The challenger sweated, he strained, he got nowhere. A jolting pain like none he'd ever felt roared through him, and he instinctively tried to release the box, but his hands stuck to it like a tongue to frozen steel. Then the pain stopped and he collapsed, humiliated but unhurt. Robert-Houdin walked over and lifted the box without a huff or a puff. Robert-Houdin 1, Marabouts 0. Long story short, the Algerians chose not to revolt, and Robert-Houdin became a national hero.

The secret to Robert-Houdin's unliftable box was an electromagnet concealed in the ground (placed there by a savvy Robert-Houdin the night before). When he switched on the current, the box became unliftable, and when he amped up the current, he gave the challenger the shock of his life. Robert-Houdin changed the Algerians' perceptions. Some who had believed the Marabouts had real powers were wised up. Others continued to believe in the Marabouts' powers but were convinced that the Frenchman's powers were greater. Robert-Houdin's genius altered the challenger's kinesthetic perception and the onlookers' visual perception, then their emotional willingness to revolt.

The Robert-Houdin story is not as quaint as it might sound at first to a 21st-century reader. When I perform in countries where the belief in magic is strong, I have to be careful to explain that I'm an entertainer and an *illusionist,* that what I do is achieved through the laws of optics and physics and misdirection, not via the paranormal. And I sometimes still run into trouble.

I remember being challenged on a number of occasions by local magic men, who thought I was there to show them up. I had to explain that what I do is very different from what they claim to do, and that I intended no offense. And yet on more than one occasion I had to have bodyguards when a local magic man refused to believe that what I do is pure entertainment and challenged me to a magic duel. Those were lessons in how cultural differences affect perception.

The kind of perception I deal with chiefly, though, is based in biology and psychology. The human brain—the most complicated organ on the planet—is the theater where the magic I perform really takes place.

The hand is not quicker than the eye, but the hand is quicker than perception. If the brain knew what to look for, the eye would see it. Tricks of attention, for example, are responsible for some of magic's greatest effects, and the ability to manipulate the audience's attention is one of the magician's most crucial skills.

If I can gather your attention and fix it on something specific, there's a very good chance that you won't notice things that are happening right in

Fact

Magicians performed in ancient Greece and Rome. In ancient Egypt one named Dedi cut off and restored animals' heads for King Khufu. Because magic needs no words, it is universal.

> What the eyes see and the ears hear, the mind believes.
>
> Harry Houdini

front of you. Suppose I take a person who's a huge baseball fan and I say, "You're about to meet [fill in the blank with the name of your all-time favorite player]," and then I bring that player over to meet the fan. I can have my assistants literally walk an elephant into the room, right in the fan's field of vision, and there's a pretty good chance he won't even notice the elephant! The person's attention is so sharply focused on the player that he doesn't perceive what the player sees! The brain can focus on only one thing at a time. It's not just your eyes that are focused; it's also your attention and your thoughts. With magic you are even trying to decipher the act itself, which causes the focus to become still sharper.

Good misdirection is mostly psychological, with the magician tricking different parts of the audience's brain. With good misdirection, the viewers don't even know that they've missed anything or been deceived; they just experience the magic. If I take an envelope and lick the flap and seal it, the viewers will assume the envelope is sealed and nothing can be slipped into it. It might be completely open on one side, but because I casually show it and close it, the assumption is that the envelope is undoctored. But if I picked it up and said, "This is an ordinary envelope, nothing fake about this," I'm casting suspicion on it by calling attention to it, by making it a part of the viewer's focus.

Science is now labeling and analyzing things that magicians have known for centuries. For instance, some of the most deceiving moments involve what scientists refer to as change blindness, as demonstrated in illusions where audience members don't notice obvious changes in their visual field when their focus is narrowed to a specific scope or task. Sometimes, the closer you look, the less you see. And that is what makes magic so fun.

Then, too, there are straight-up optical illusions that deceive the eye and, therefore, the brain. Hundreds of years ago, magicians discovered, for example, that if a stage is draped in black, anything on the stage that's also black can't be seen by the naked eye. This principle, which magicians call black art, delighted and perplexed me as a kid when I first encountered it in the form of a mouse with an Italian accent on *The Ed Sullivan Show.* Topo Gigio, the mouse puppet, was unlike any puppet I'd seen: He had no visible means of support. He stood on his own two paws and often crawled up Sullivan's sleeve to give him a good-night peck on the cheek. Topo Gigio was like a cartoon come to life.

What I didn't know, and wouldn't learn until I checked a book on magic out of the library in the sixth grade, is that Topo did have handlers, but they were invisible because they were dressed entirely in black, with black hoods and black gloves. Topo was brought to life by puppeteers in plain view and yet completely invisible to the camera and the studio audience.

Optical illusions like that are well understood. But one of the most fascinating features of this book and the companion television special, *Brain Games,* is their exploration of illusions and brain processes that magicians have known and exploited but never completely understood. Reading these explanations of why a certain perceptual manipulation works has deepened my appreciation for what we illusionists do and sharpened my use of the tools we keep in our toolbox.

As an illusionist, I help people recapture their sense of wonder by creating amazing things they've never seen before—what actors call the illusion of the first time. Except, for my audience, it's no illusion. It's a real feeling of awe and raw astonishment. A sense of enchantment—that's what so many of us are missing, particularly now that we have so much wonderful technology at our fingertips. We can create near miracles with our laptops and our tablets and our smart phones. When I can unplug the audience for an hour or two and give them back that sense of total enchantment, it's the greatest feeling. It's the reason why I became an illusionist, and it's what gets me on stage day after day, year after year.

This book is an extraordinarily powerful and fun tool for enriching your knowledge of perception and capacity to wonder. You will learn not only to look closer but to see and experience more. I'm honored to be a part of this project, which confirmed many things I had come to know through my work but couldn't quite articulate and which taught me things that are both useful and entirely fascinating. I'm delighted to be on this journey with you.

Topo Gigio

American variety show host Ed Sullivan strikes up a conversation with puppet Topo Gigio on *The Ed Sullivan Show,* October 4, 1964.

YOUR BRAIN

Today's fastest supercomputers can perform millions of mathematical calculations within a single second. They can send messages from person to person around the world, adjust the flight of rockets zipping at bulletlike speed to intercept other rockets, and checkmate grand masters at chess without breaking an electronic sweat. Yet no machine available today comes close to matching the computational ability of the human brain. Machines are not poets, architects, doctors, or artists. They do not think. And, perhaps surprisingly, they have great difficulty making even the most rudimentary sense of the world. • The brain makes humans unique. While it duplicates many of the functions of other animal brains—including the analysis of stimuli from the five senses; the coordination of muscle movement; and the regulation of heart, lungs, and other organs—the human brain also creates consciousness. Human brains synthesize and internalize a version of the world and take the added step of creating awareness of one's place in that world. Unlike animals, humans know that they know. And they can choose how to act in response to that knowledge.

YOUR SELF

Out of billions of cells...

Neuron

Neurons—the brain's building blocks—receive information through dendrites and then forward it via axons to spur actions from simple to sublime.

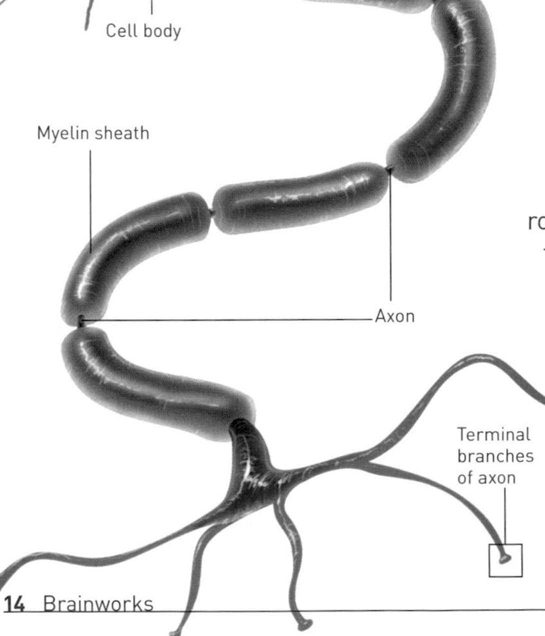

The brain's basic unit is the neuron. It is a specialized cell designed to share information electrochemically with other neurons. Some chains of neurons send information to the brain from the body's extremities. They tell the brain to register the pain of a finger struck by a hammer, the noise of passing traffic as it falls upon the ears, and the sublime colors of an Arizona sunset. Other chains send information from the brain to the body. They direct fingers to type, tongues and lips to form words, and eyes to swing right and left to focus on the ball at a tennis match. Other chains share data among themselves to construct subconscious or conscious thoughts and feelings.

Each neuron contains a cell body with a long, tail-like fiber called an axon. The axon's job is to send electrical impulses to other cells, thereby telling muscle cells to contract, relaying sensations from the body, and otherwise sharing information with other neurons. Some axons are short, extending only to adjacent cells in the brain. Others are much longer, carrying impulses down the spinal cord to move the arms, legs, and feet. Axons may split and branch into as many as 10,000 knoblike endings that disperse impulses across many cells.

Each neuron also extends into networks of dendrites, which are thin, short fibers that transport electrical signals to the main body of the neuron from outside sources. Under a microscope, axons and dendrites somewhat resemble the roots and branches that form the myriad tangles of a mangrove swamp. Only in this case, the branches of one tree reach out toward the roots of another, and vice versa.

The human brain contains perhaps 100 billion neurons. Each neuron links to so many others that the entire network forges literally trillions of connections, making the brain the most complicated object in the universe. And because humans and other animals can learn, these connections do not remain static. Every day, as the brain incorporates new experiences and new knowledge, neurons forge new connections. They can

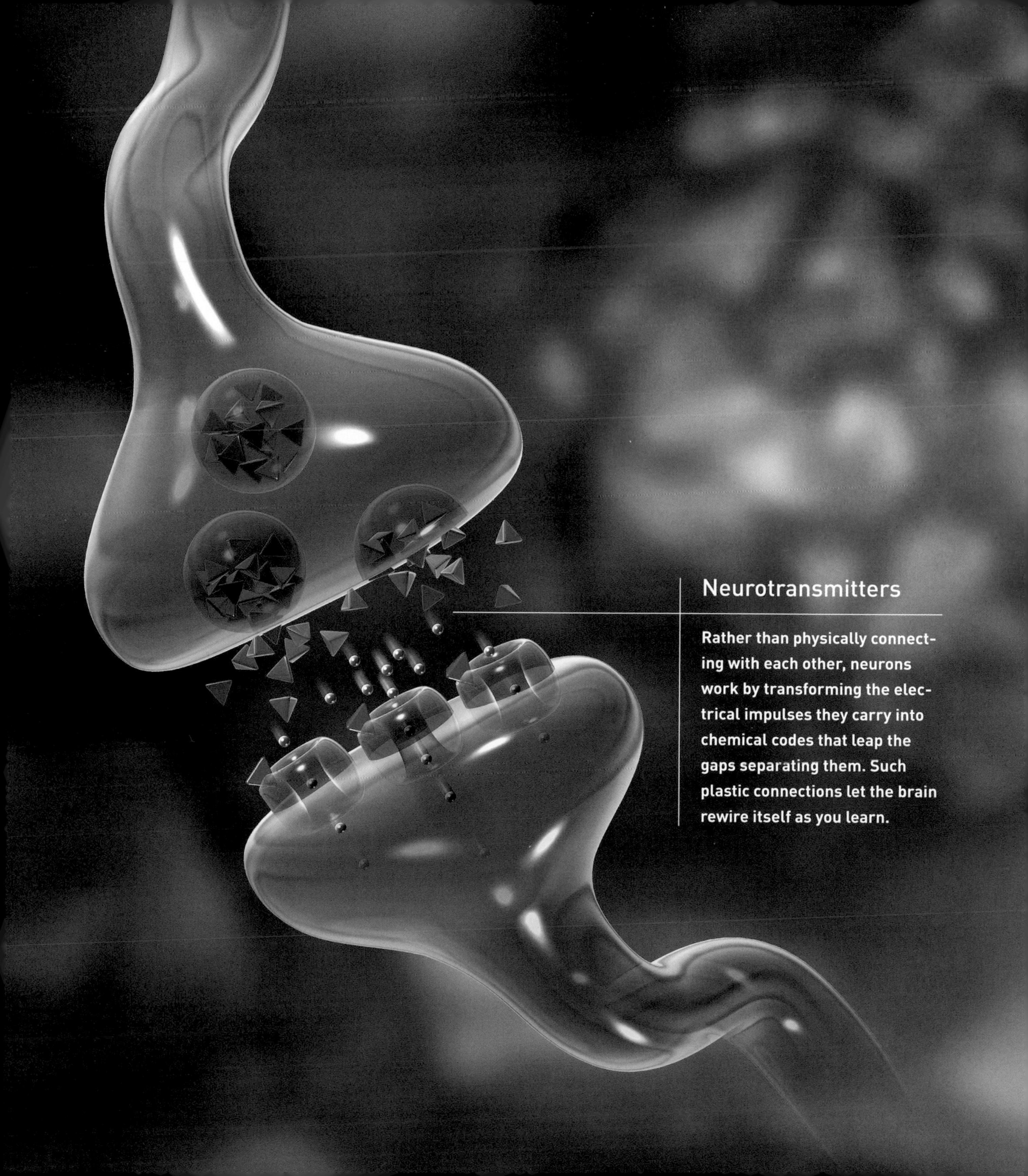

Neurotransmitters

Rather than physically connecting with each other, neurons work by transforming the electrical impulses they carry into chemical codes that leap the gaps separating them. Such plastic connections let the brain rewire itself as you learn.

...the brain forms...

Action potential

A motor neuron sends branching fibers onto a muscle cell's surface. When active, the neuron communicates electrochemically with cell receptors, and this causes the muscle to contract.

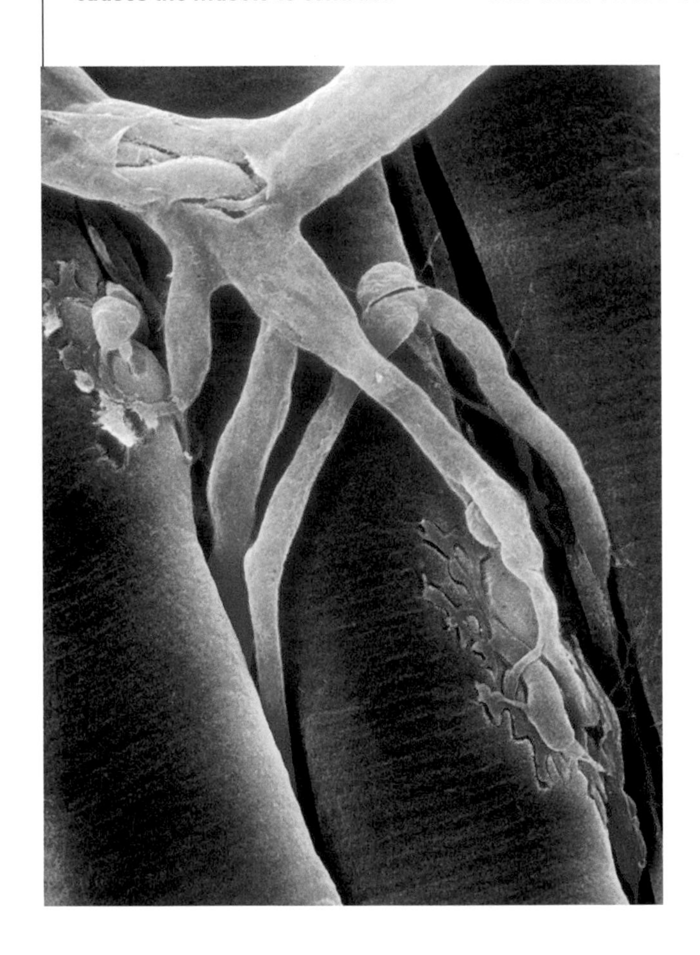

do so because neurons do not join one another like bricks mortared together in a wall or steel girders welded to form a bridge. Instead, a small gap called a synapse lies between the axon of one neuron and the dendrite of another.

When a neuron sends an electrical discharge along the length of its axon, it halts at the synapse like a car at the edge of a cliff. There, the impulse activates electrically charged molecules stored in the neuron's cell wall. These molecules, known as neurotransmitters, leave the membrane of the first neuron, move across the synaptic cleft, and dock at a second neuron. The arrival of a neurotransmitter alters the electric charge at the edge of the new neuron and sparks a new electrical impulse. As impulses pass among complicated chains in the central nervous system, they form networks that specialize in performing particular functions, such as understanding language, remembering experiences from the past, and comprehending the outside world. They store, retrieve, and transmit information. All information processed by the brain is nothing more than electricity passing through neuron after neuron and pausing only to be converted into chemical energy as it leaps across each synapse.

Neural networks lie in four main parts of the brain. By evolutionary reckoning, the oldest portion of the brain is its stem, which begins as an extension of the spinal cord. The brain stem controls basic physical actions necessary for survival, such as heartbeat and respiration. It is home to many sensory and motor nerves, the latter named for their function of controlling movement in muscle tissue. Motor neurons also densely populate a second part

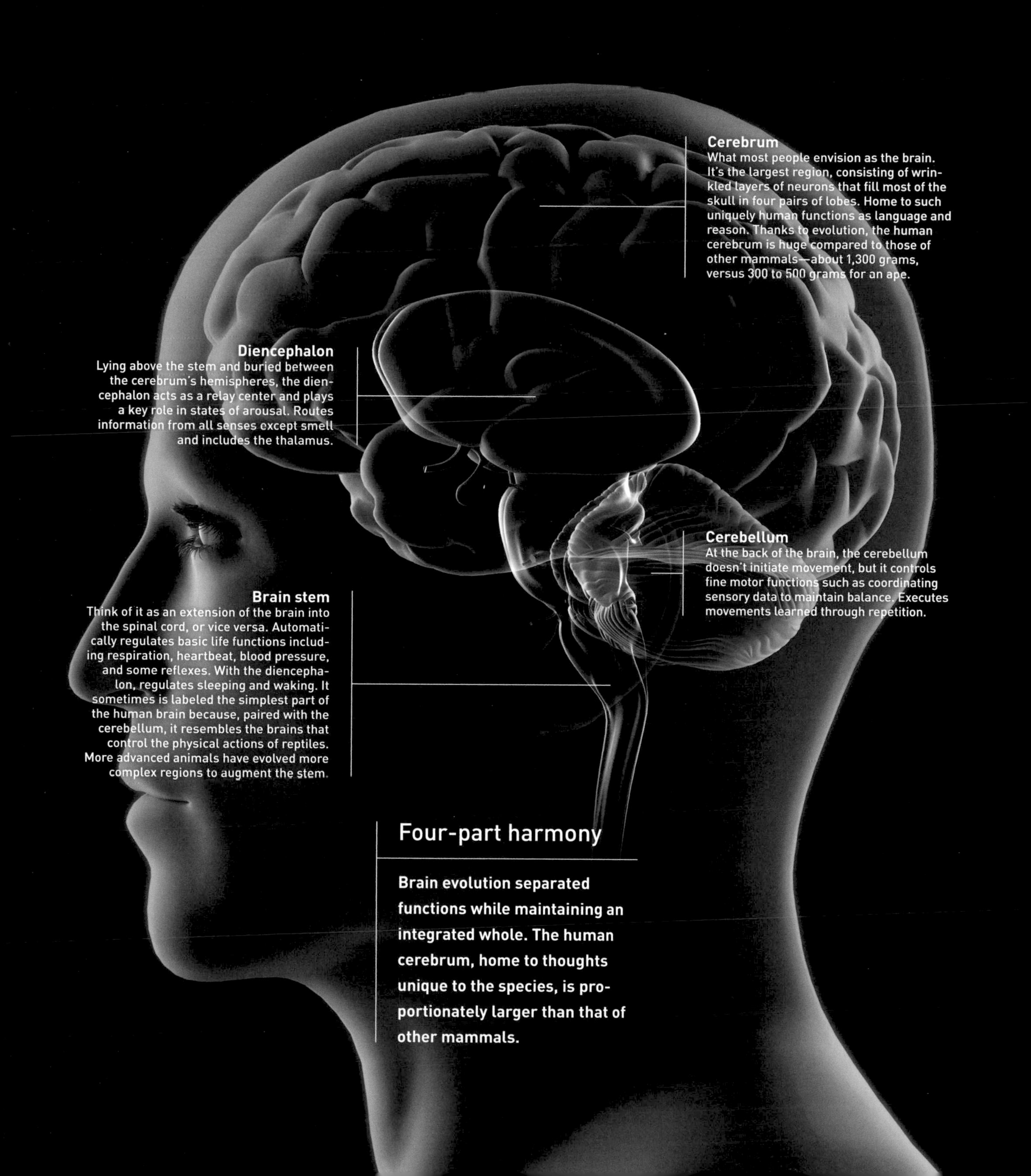

Four-part harmony

Brain evolution separated functions while maintaining an integrated whole. The human cerebrum, home to thoughts unique to the species, is proportionately larger than that of other mammals.

Cerebrum
What most people envision as the brain. It's the largest region, consisting of wrinkled layers of neurons that fill most of the skull in four pairs of lobes. Home to such uniquely human functions as language and reason. Thanks to evolution, the human cerebrum is huge compared to those of other mammals—about 1,300 grams, versus 300 to 500 grams for an ape.

Diencephalon
Lying above the stem and buried between the cerebrum's hemispheres, the diencephalon acts as a relay center and plays a key role in states of arousal. Routes information from all senses except smell and includes the thalamus.

Cerebellum
At the back of the brain, the cerebellum doesn't initiate movement, but it controls fine motor functions such as coordinating sensory data to maintain balance. Executes movements learned through repetition.

Brain stem
Think of it as an extension of the brain into the spinal cord, or vice versa. Automatically regulates basic life functions including respiration, heartbeat, blood pressure, and some reflexes. With the diencephalon, regulates sleeping and waking. It sometimes is labeled the simplest part of the human brain because, paired with the cerebellum, it resembles the brains that control the physical actions of reptiles. More advanced animals have evolved more complex regions to augment the stem.

...the complexity of you.

Cerebral lobes

Brain regions exist in left-right pairs. Frontal lobes (red) handle high-level mental functions. Parietal lobes (purple) process sensory information. Temporal lobes (green) are key to speech and long-term memory. Occipital lobes (orange) contain the visual cortex.

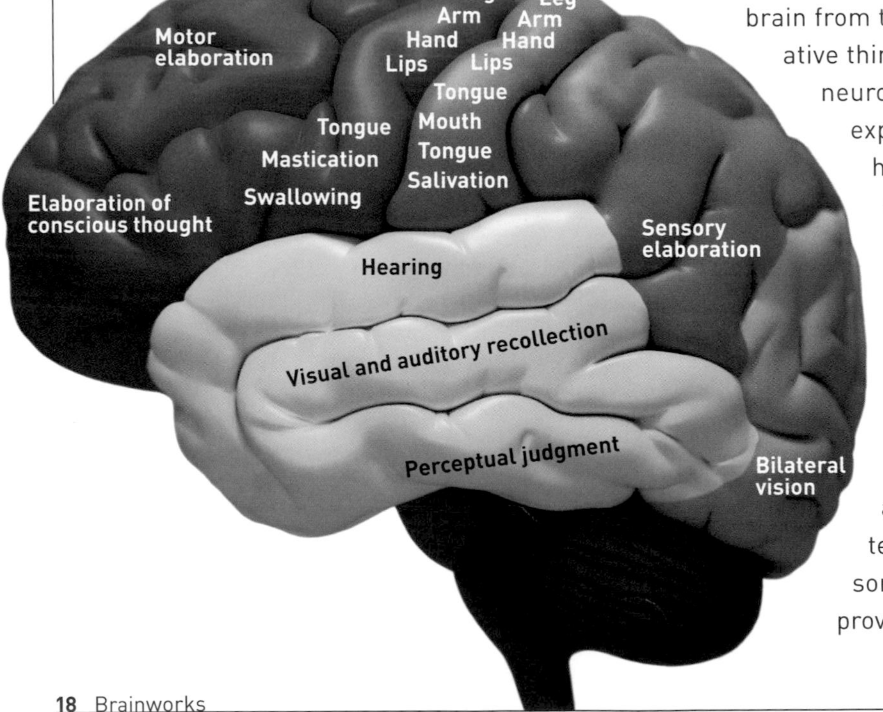

of the brain, the cerebellum, at the back and bottom of the skull. The cerebellum coordinates precise, voluntary movements, such as tying a shoe or playing a violin, and also plays a role in emotion. A third component, known as the diencephalon, lies in the brain's center. It controls the body's rhythms of sleeping and wakefulness, regulates involuntary actions of the nervous system such as digestion, and relays sensory stimuli to other brain regions.

The fourth region, the cerebrum, is what most people think of when they envision the brain. It lies in two hemispheres, left and right, connected by a band of tissue called the corpus callosum. The cerebral cortex is the outermost brain layer, folded and wrinkled and resembling a squishy pink walnut. Neuroscientist Michael Gazzaniga likens the cortex in size and appearance to a large dish towel folded and laid atop the rest of the brain. The cortex is home to the higher functions that separate the human brain from those of other animals: reason, creative thinking, and language. The amount of neurological firepower necessary for such exploits is considerable: 76 percent of human brain mass lies in the cerebral cortex, a greater percentage than that of any other animal, and within the cortex lies about 75 percent of all neural connections.

The cerebral cortex processes information so that you may comprehend enough about the world to survive—and even to thrive. Evolution and experience have molded the cortex's neural connections to favor sensory and cognitive functions that have proved successful over eons of human life.

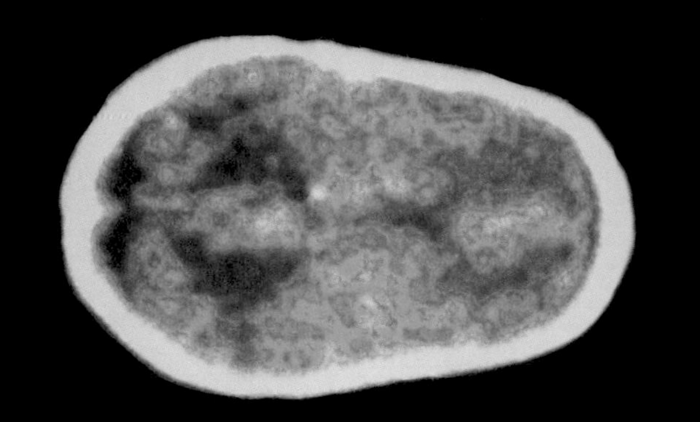

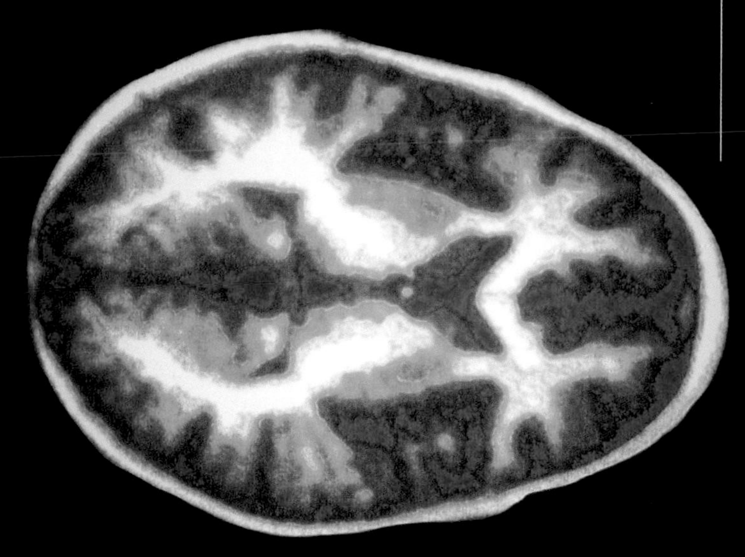

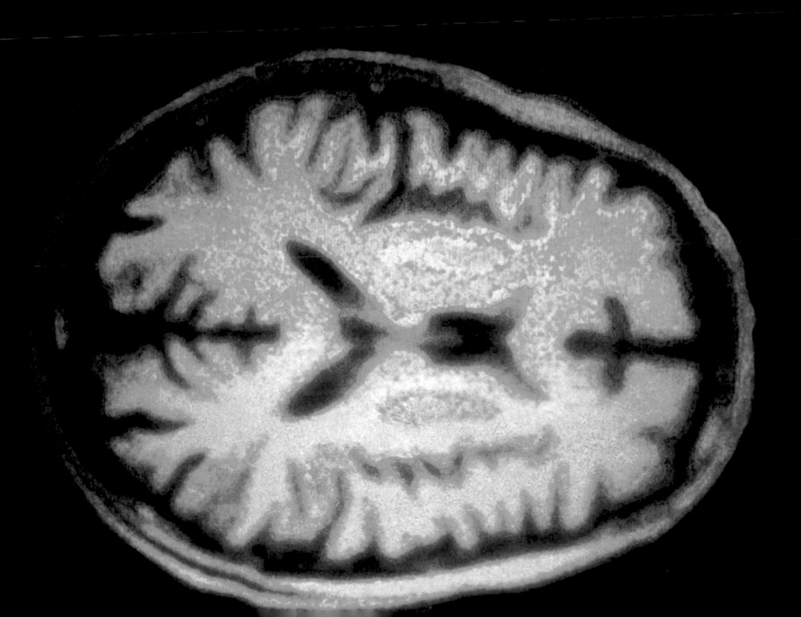

Brain development

A newborn's cerebral cortex (top) shows little evidence of folding. By age five (center), a child's brain is the size of an adult's (bottom) but is still undergoing folding, which maximizes cortical surface area. Unfolded, an adult cortex would cover 2.5 square feet.

Building up

Mammalian development of a limbic system added emotions and more complex behavior. The prefrontal cortex introduces the highest mental functions and can override emotion.

Anything that challenges your brain's time-tested circuitry, such as the illusions and paradoxes of this book, opens a window onto hidden depths of self-knowledge.

Your brain's ability to interpret what it experiences adds complexity to the end product of evolution that lies within your skull. A human brain, which can ask questions about what it sees and knows and then ponder what's gained by the answers, must turn to the ultimate question: Just who is it, posing and solving these problems? A journey into your brain leads to yourself.

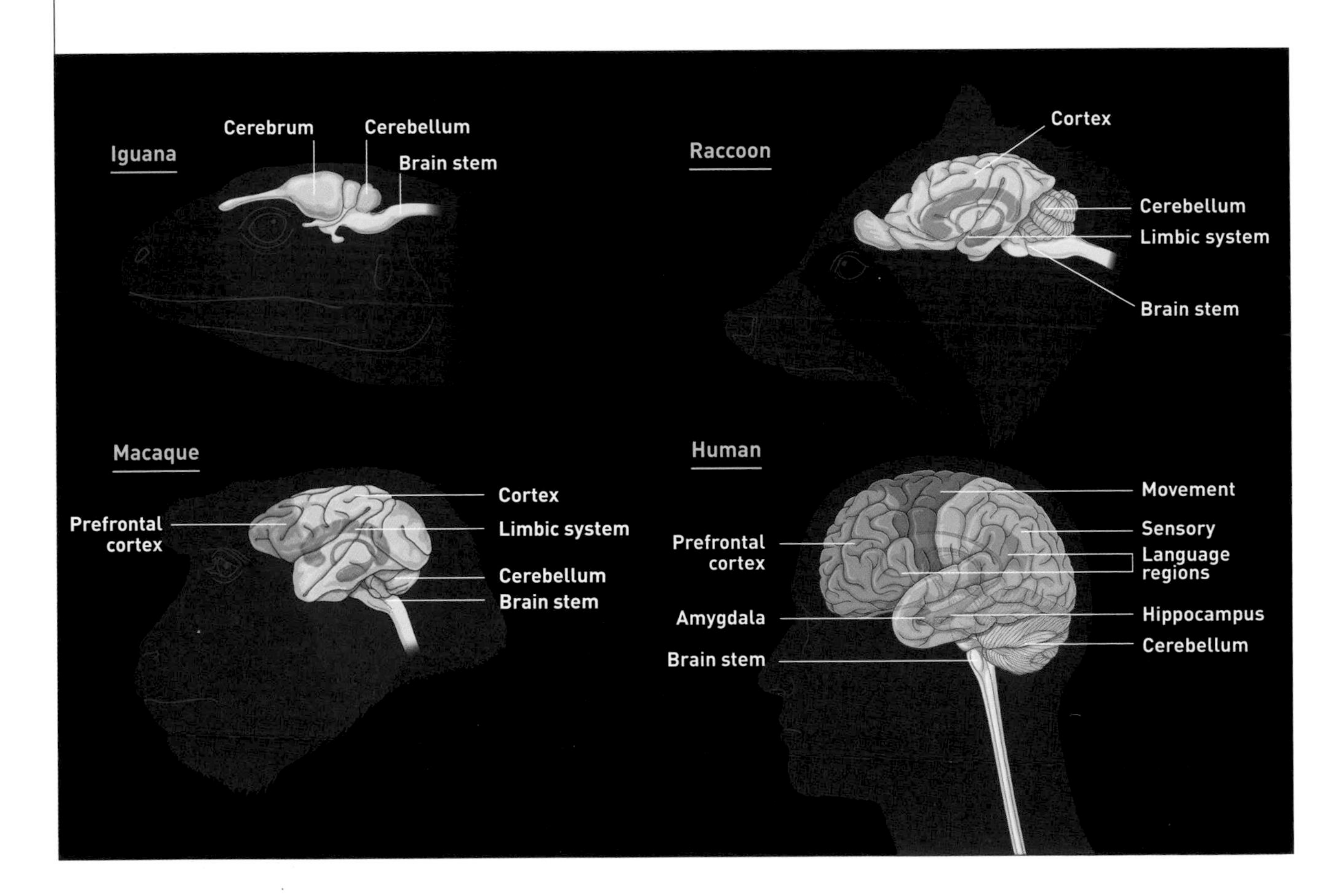

How to use this book

Your brain is the epitome of synergy, the phenomenon of the whole being greater than the sum of its parts. So, too, is *Brainworks,* a companion book to the National Geographic Channel special *Brain Games.*

Each National Geographic project can be enjoyed on its own merits; the two projects, television series and book, intersect at key points but do not duplicate. You'll learn much from either, but exposure to both weaves the richest tapestry of knowledge.

This book has a general introduction and three chapters: **Seeing,** illusions of vision and what they reveal about the brain; **Thinking,** phenomena of thought and memory; and **Being,** mental exercises that illuminate consciousness, emotion, and will.

Each chapter begins with an introductory essay that outlines something of the history and current understanding of the issues at hand. Occasionally, a **fact** (called out by an arrow) or a **quotation** (quotation mark) punctuates the text by underscoring key concepts. These introductions set the stage for what follows: **Experiments** of sensation and thought that you can do, just as if you were in a neuroscientist's laboratory. Each experiment is set up by brief instructions, contained to the right-hand page, and an accompanying illustration. You are encouraged to carry out the experiment's instructions before turning to the next page.

From there, you will read **"What happened"** (question mark), a quick word of explanation. **"Look closer"** (exclamation mark) takes you deeper into the neurological basis for the phenomenon under study. **"The takeaway"** (asterisk) summarizes the experience and its significance in plain language. Many experiments are supplemented by eye-opening case studies.

Chapter 1
SEE

I N G

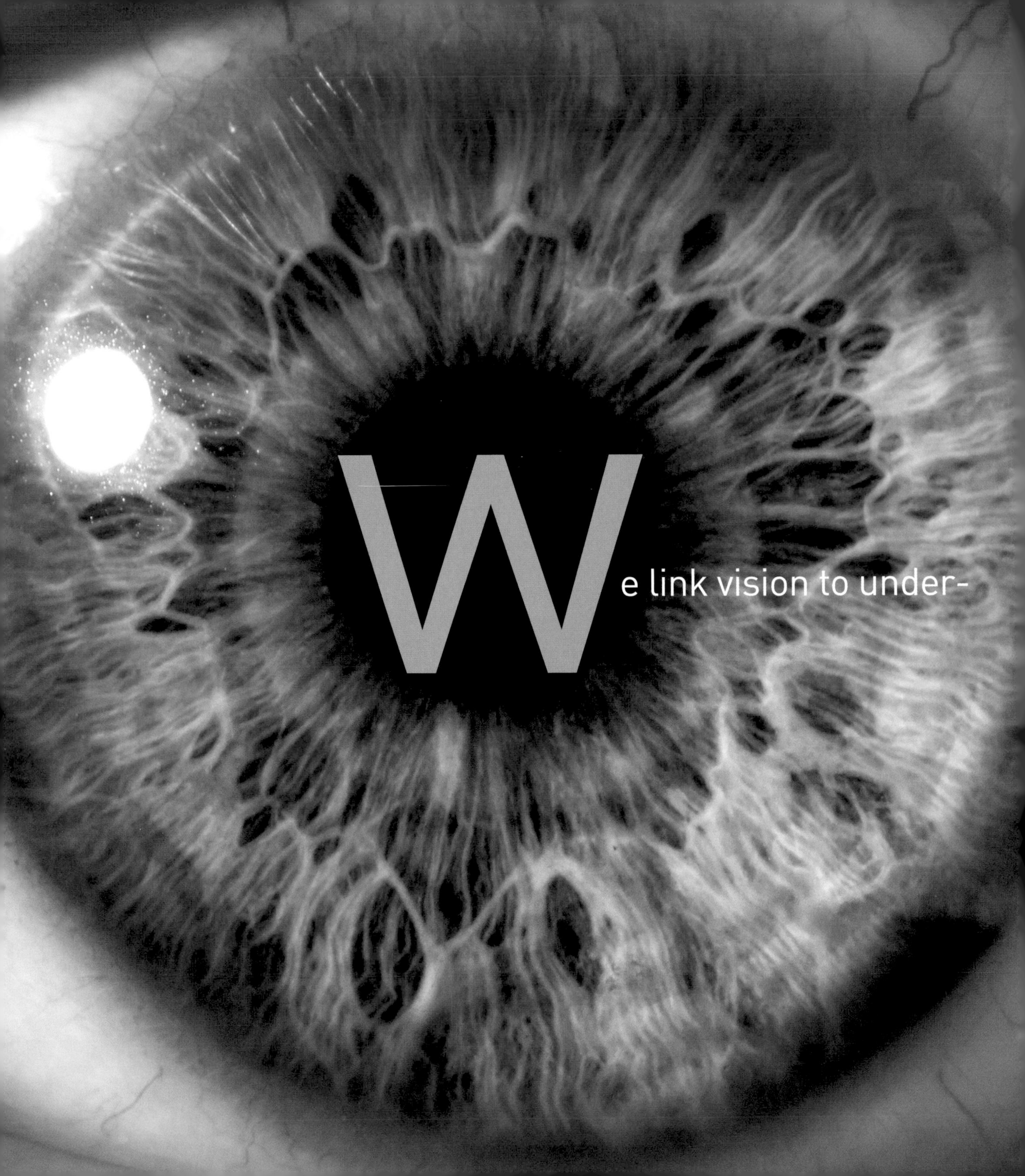
We link vision to under-

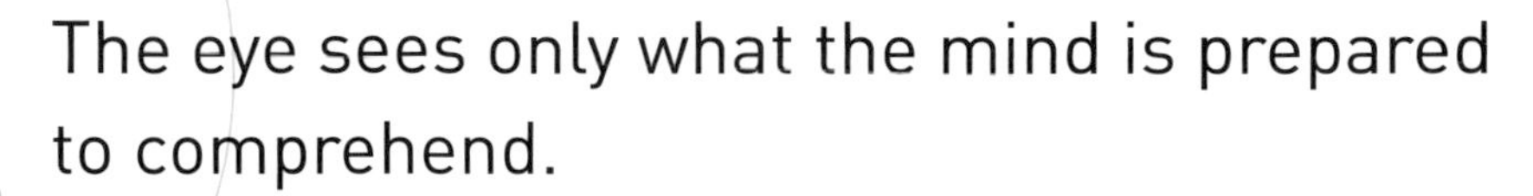

The eye sees only what the mind is prepared to comprehend.

Henri-Louis Bergson

standing. Language tells us so. "I'm in the dark," we say when a point remains unclear. "I see!" we exclaim at the moment of clarity. "Show me," says the skeptic from Missouri. We trust in vision. We consider what we see to be a true representation of the external world. But the brain, working with the nerve cells of the eyes, constructs the outside world inside our heads. And the brain and eyes do this actively, not passively like some pinhole camera. We see what the brain tells us we see. The process is mind-bogglingly complicated. • Reduction of the three-dimensional world into the two-dimensional simulacra of vision begins with the eyeball. Beams of light reflected from objects enter the eye through the pupil and are bent by the cornea and lens. The cornea is the transparent covering atop the iris, pupil, and interior of the eye. Its shape remains relatively constant, but its curvature helps gather

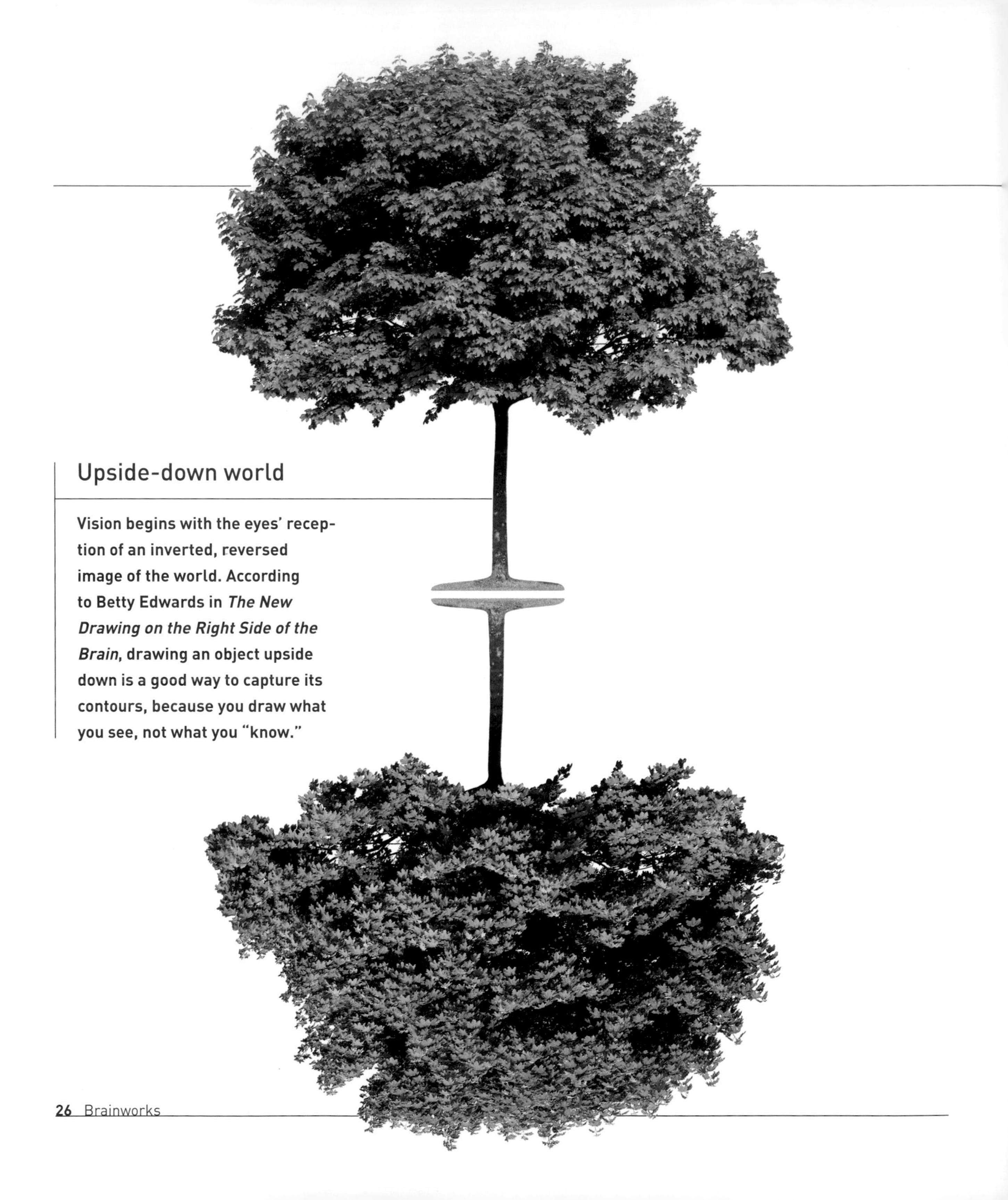

Upside-down world

Vision begins with the eyes' reception of an inverted, reversed image of the world. According to Betty Edwards in *The New Drawing on the Right Side of the Brain*, drawing an object upside down is a good way to capture its contours, because you draw what you see, not what you "know."

light like the curved front of an old-fashioned camera lens. The eye's lens, just behind the iris and pupil, normally is a clear, flexible structure, although it can become cloudy or rigid with age or disease. It changes shape, allowing the eye to focus on objects at different distances. Light waves from distant objects move in nearly parallel lines when they strike the front of the eyeball, while light waves from nearer objects arrive at greater angles. Both cornea and lens are denser than air, so they *refract*—change the direction of—any light beams that reach them through the medium of Earth's atmosphere. The curvature and density of the cornea and lens redirect and focus incoming light waves at the back of the eyeball to register on the retina.

Fact

Every retina has a blind spot where the optic nerve exits to the brain. French physicist Edme Mariotte discovered it in 1662.

A magical membrane

The retina is an extension of the brain. It forms as a pocket of the developing brain of an embryo in the uterus and grows into an astonishingly complex membrane of cells that have evolved to detect a narrow band of electromagnetic energy: visible light. The brain's visual understanding of the world begins when

Inside the eye

Light reflected off an object enters the eye through the clear cornea and then passes through the pupil, the circle in the center of the iris. Light waves are focused and redirected by both the cornea's curves and the lens (a shape-shifting structure behind the pupil and iris) and sent to the retina, a light-sensitive extension of the brain.

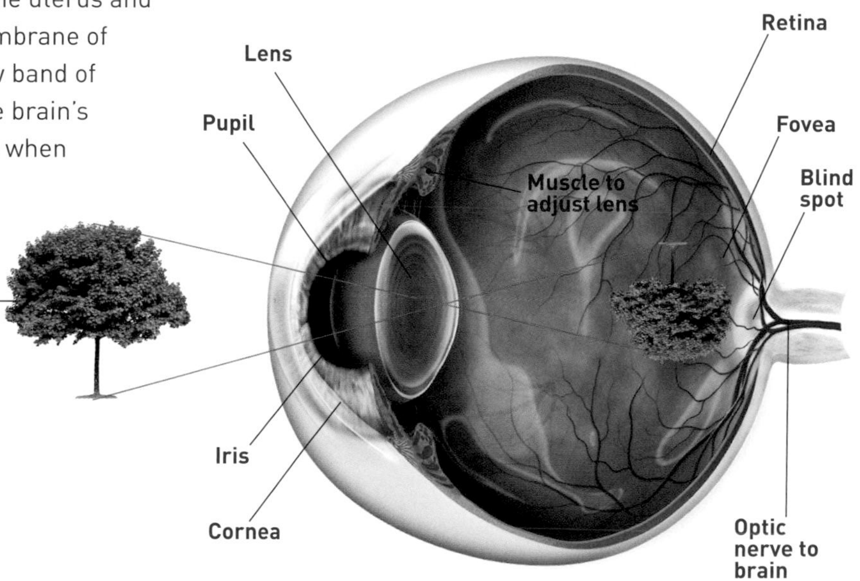

The eye, to this day, gives me a cold shudder, but . . . my reason

Rods and cones

A colorized scanning electron micrograph reveals the structure of the retina. At bottom are nerve fibers that form the optic nerve. It receives signals from rods (green), which are long nerve cells that respond to dim light, and cones (pink), the shorter, less sensitive cells that detect colors.

electromagnetic energy traveling in wavelengths between about 400 and 700 nanometers (*nano-* = "billionth"), seen as ranging from violet to red, gets absorbed by highly specialized pigment molecules embedded in the retina's photoreceptor cells.

Photoreceptors are capable of detecting a candle in the darkness two miles away and functioning when light is millions of times brighter on a snowy day. These tiny superhero neurons of your eyes tell your brain how much light is located where. Once light bleaches the pigment in your photoreceptors, much like the sun-sensitive paper you may have played with as a child, the photoreceptors send tiny bursts of electricity to nearby, connected neurons. After going through four other levels of neurons, all located in your retina (remember, the retina is part of the brain), the visual information exits the back of the eye and gets sent to other parts of your brain. This cascade of electricity changes the light in the world to impulses traveling through your brain.

Although all photoreceptors use similar mechanisms that react to light and transmit information, you have two different types, making up two systems in your eye.

Rods and cones

Your nighttime visual system, made up of receptors called rods, is active in low-light environments, such as a crescent moon and a scattering of stars on a partly cloudy night. Extremely sensitive to light, the pigment inside the rod photoreceptors changes its activity in response to a single photon. But it is not as sensitive to the exact location of the light (visual acuity, or focus) or what kind of light it is (wavelength, or color). Rods don't excel at visual acuity because they are fairly evenly distributed across your retinas (although absent in the very center of your vision), and each rod is sensitive to a relatively large area of the world. Further, your retinas' bipolar and ganglion cells must aggregate information from many rods. So when a ganglion cell, the last level of neuron in the retina, gets information from the rods, it cannot be sure of the message's point of origin on the retina. Rods aren't sensitive to color because there is only one kind of rod, and while they are more sensitive to certain wavelengths—ever notice that blue lights look especially bright at night?—they are not good at distinguishing wavelengths, an ability necessary for color perception.

Our other system, made up of cone photoreceptors, provides daytime (high-illumination) vision, color perception, and acuity in central vision—in other words, all of the ingredients you need to read this book. Cones come in three varieties, each with a distinctive photosensitive pigment. These cone cells are classified as S, M, and L, for short, medium, and long. Short-wavelength

Retinal cross section

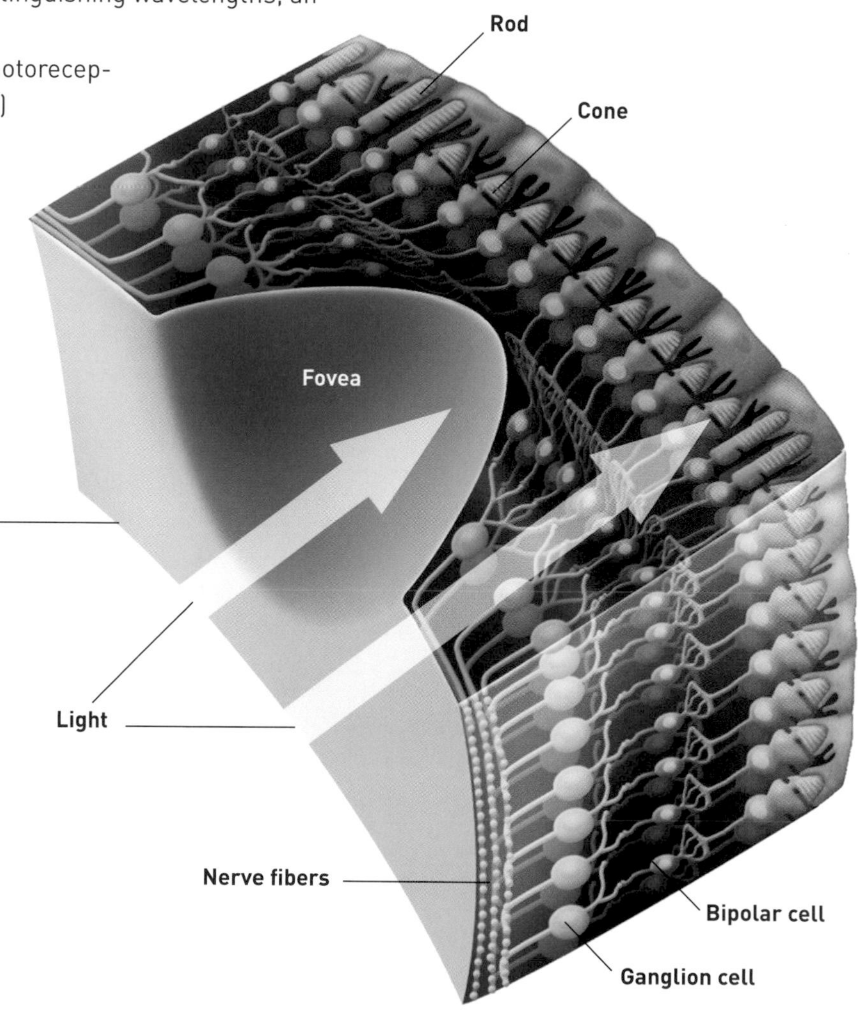

About 7 million cones and 120 million rods cover the retina on the eyeball's inside back lining. They connect to a middle layer of bipolar cells, which link to inner ganglions, whose axons form the optic nerve. Also in the retina lie connective amacrine and horizontal cells. Light, indicated by large white arrows, passes through other cells to reach the rods and cones.

Fact

If one photon strikes a rod, the rod will fire in response, but six to eight photons must strike before you consciously notice the change.

light stimulates the S cones to initiate the perception of color roughly in the blue range. Medium-wavelength light leads to perceptions roughly in the green range in the M cones, and long-wavelength light stimulates perception roughly in the red and yellow range in the L cones. The combination of electrical stimuli from all three kinds of cones results in the brain's recognition of all the familiar colors.

Cone-based vision is much sharper than rod-based vision. That's because only one or a very few cones converge their nerve impulses into a bipolar cell. It's also because cone cells are packed tightly in the center of your retina, called the fovea. When you get to the end of this sentence, focus on the final word without moving your eyes. Did you notice how the word *eyes* was sharp and clear, but three or four words before and after it were not? As you rested your eyes on *eyes*, its image fell on the tightly packed cones in your fovea.

Pathways of vision

Much synthesis of visual information occurs in the retina, as signals from cones and rods get processed by three other cell types: bipolar, horizontal, and amacrine. We don't entirely understand the roles these cells play as intermediaries between the rods and cones and the visual cortex of the brain, but their connections suggest an increasing complexity of visual information as it passes into the ganglion cells, whose mass of axon fibers forms the optic nerve.

The two optic nerves—one for each eye—relay visual information to each half of the brain, depending on which side of the retina the information is coming from. Information from the left sides of the two retinas, which observe objects in the right-hand side of the field of vision, gets routed to the left hemisphere. Information from the right sides of the retinas, which observe objects in the left-hand side of the field of vision, gets routed to the right hemisphere. So, your right brain sees the left side of space, and vice versa.

Besides being split in half that way, the information from each eye is also split between two places in each half of the brain. Most goes first

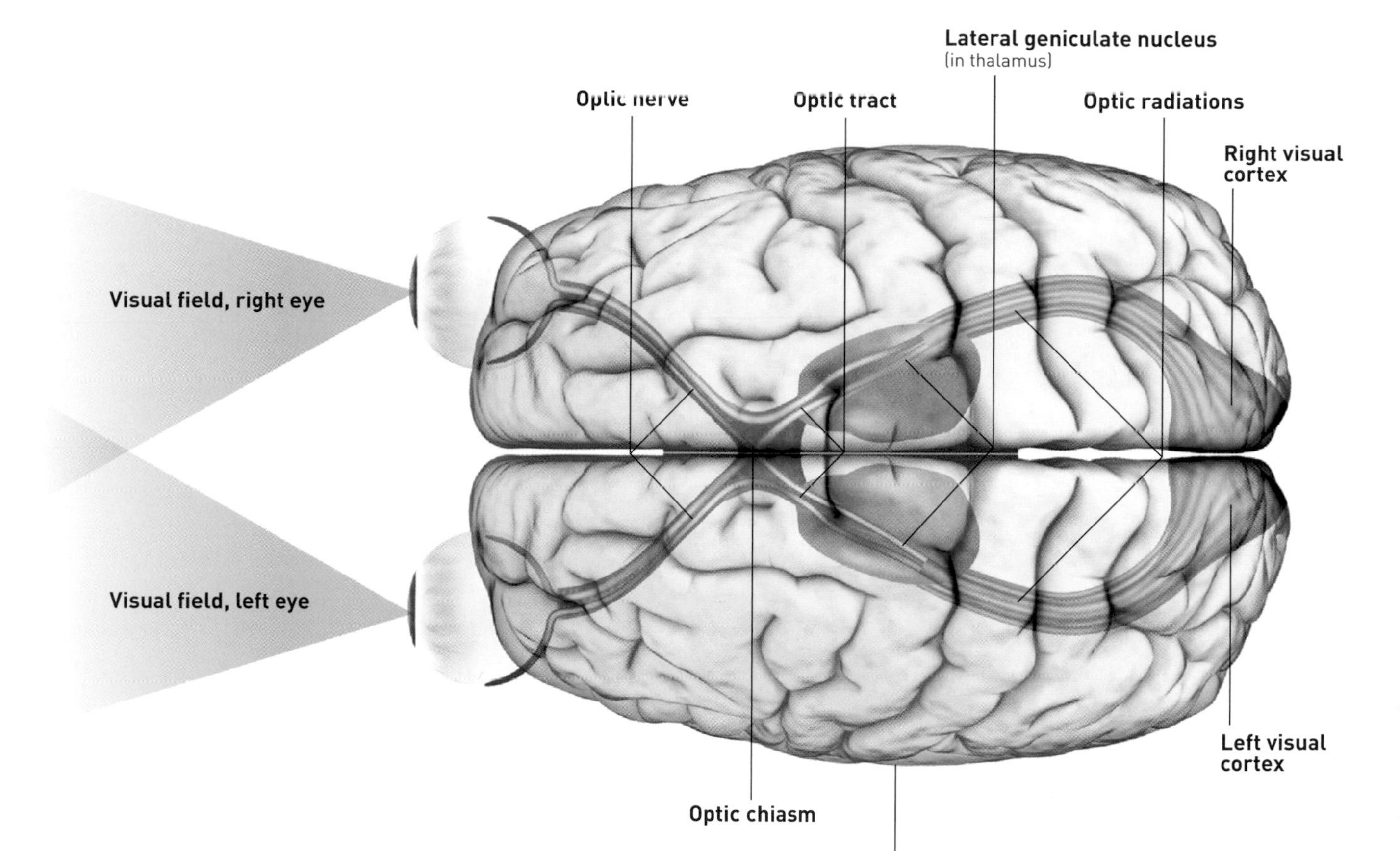

to the thalamus, a sensory command center tucked in the middle of the brain just above where your spinal cord meets your brain. However, some information goes to the superior colliculus, a structure that sits below the thalamus and is responsible for orienting your eyes when they detect a moving object in the periphery, such as a fly about to buzz by your head. The superior colliculus sends a command to your motor cortex—"Quick, move your head!"—while the thalamus (or lateral geniculate nucleus of the thalamus, to be exact) sends information to the visual cortex in the occipital lobe at the back of your head.

The primary visual cortex is also known as the striate cortex—because of its distinctive stripes (striae)—or the V1 cortical area. Visual processing begun there moves to areas V2 through V5 but then diverges again into two more visual streams. One goes up, toward the parietal lobe at the top of your head and the sensory and motor cortices, and the

Twin circuits

Each optic nerve splits the information it receives and routes some to each half of the brain. The right half of your brain sees the left side of your field of vision (which strikes the right side of your two retinas), and the left half of your brain sees the field of vision's right side. Visual information goes to the thalamus, just above the spinal cord, and is relayed to the visual cortex in the occipital lobe for further processing.

> Your brain . . . takes in your world like a huge flood of data and acts like an irrigation system. Jonah Lehrer

other goes along the temporal lobe, behind your ears, near your language areas. The dorsal stream is sometimes called the "where" or the "how" pathway because tests have linked it to the perception of motion, location, and visually guided action. The ventral stream is the "what" pathway, related to object recognition.

Visual hierarchy

The brain contains many higher-order visual centers in many regions. Working together, they assemble the bits of electrochemical energy that begin as the firing of individual neurons in the retinas. Instead of simply recognizing a projected image like the kind that strikes the film in an old-fashioned camera, the brain processes vision by breaking it into millions of bits that encode such data as color, line, shape, intensity, and motion, and then reassembling it through many neural pathways working in parallel. The quick flash of raw retinal information routed to the cerebral cortex from the thalamus serves to provide a logical blueprint for recognizing the product of the assembled information. Thus, while the visual cortex works to manufacture an image out of bits of raw

Where and what

The occipital lobe forwards visual data for analysis. According to the two streams theory, a dorsal stream goes to the parietal lobe. This channel is crucial for detecting motion, locating objects, and guiding actions. A ventral stream to the temporal lobe, near areas devoted to long-term memory, helps you recognize objects.

Where is it?
Analyzing the "blivet," or impossible object, at left, the brain's "where" pathway (red arrows) struggles to orient the four columns.

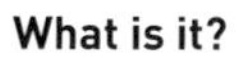

What is it?
The "what" pathway (orange arrows) identifies classical architecture apparently spanning a pool of water but still realizes something is wrong.

> The human brain, then, is the most complicated organization of matter that we know. Isaac Asimov

data, the brain's higher functions already have reached a hasty conclusion about the object most likely to emerge from the processing. Neuroscientists refer to this as a combination of bottom-up and top-down processing.

When one or more of the neural circuits that carry visual information suffers damage, the results can be quite revealing. Scientists have determined the roles of many brain regions by examining patients who have had brain injuries. Localized neuron deaths, such as those suffered in a stroke, have helped researchers pinpoint regions crucial to recognizing color, motion, faces, and letters.

Likewise, optical illusions open a window into how your brain works. Because your eyes are not cameras, and what you "see" is a construct of your brain, illusions that do not match a mechanical, camera-like recording of the visual world can be analyzed for the interesting lies they tell. ■

Processing paths

The optic nerves transport signals to the lateral geniculate nucleus (center). Nerve fibers forward data to the V1 region of the visual cortex (yellow) at the back of the occipital lobe (right), which begins to process color, shape, and motion. The V2, V3, and other regions lie near the V1 region.

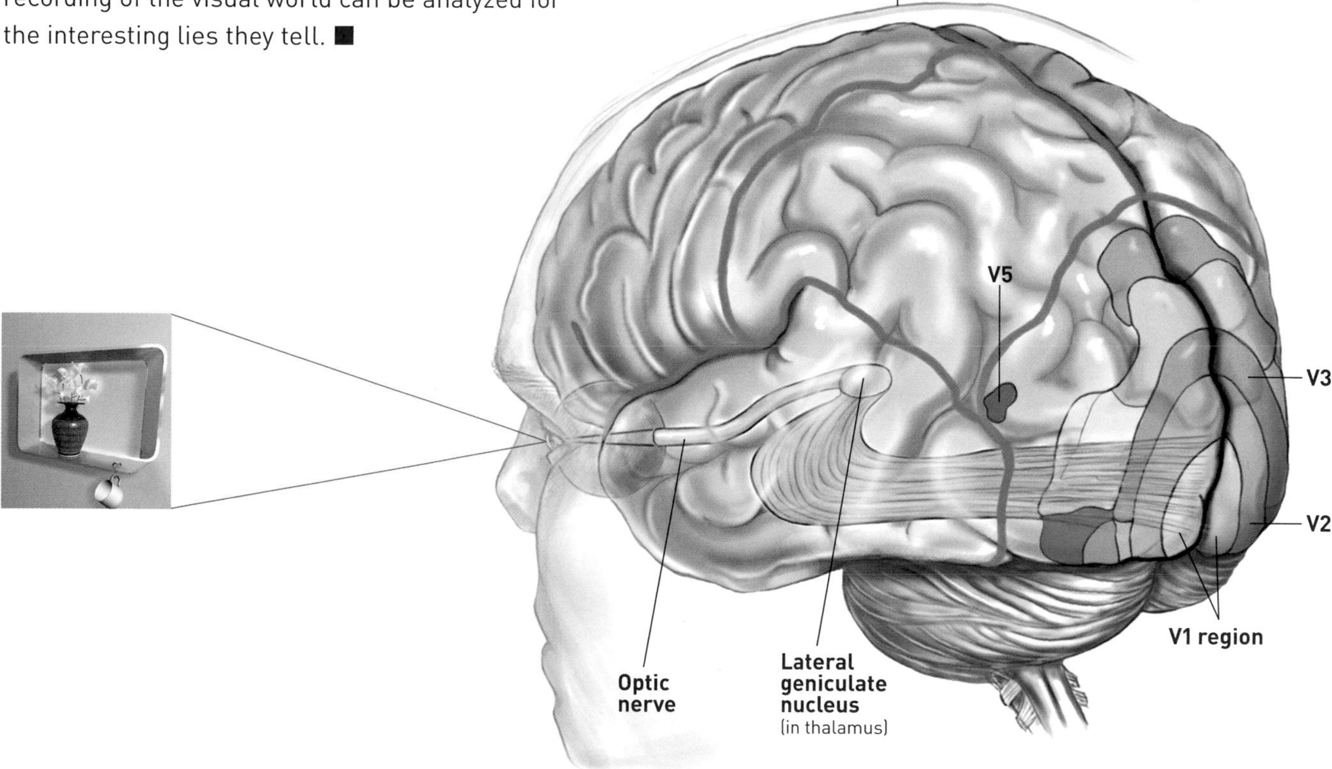

Pay close attention

to the dimensions of the two tabletops.

Ask yourself,

which would be easier to fit in a narrow dining room?

Which would be a better fit for a wide tablecloth?

In other words,

Compare

the two tabletops and decide if one is wider than the other.

Now turn the page.

What happened

The tops of the two tables are exactly the same size and shape. If you doubt this is so, measure them with a ruler.

Two common mental misinterpretations are at work in this illusion: a horizontal/vertical illusion and an illusion of three-dimensional perspective.

The horizontal/vertical illusion dates to its description in German physiologist Adolf Fick's 1851 doctoral thesis. He demonstrated differences among simple geometrical properties and how they are perceived. These kinds of disparities are called geometrical-optical illusions. Fick observed that a vertical line looks longer than a horizontal line of the same length. This is easily seen in the letter T when the horizontal and vertical strokes have precisely the same length, or when two line segments of exactly the same dimension form a right angle with one segment horizontal and the other vertical. Magazines popularized this type of illusion in the 1890s with a variation showing a man wearing a top hat. The hat looks taller than it is wide, but measuring it with a ruler proves the two dimensions are equal.

The illusion on the previous page provides a 20th-century version of Fick's T and the gentleman's top hat. The table oriented so its longest dimension appears vertically on the page is perceived as longer than the one rotated 90 degrees.

Another explanation rests on an illusion of perspective. The brain chooses to interpret the drawing as two tables. Applying the rules of perspective that it has formed through experience, the brain views the table on the left as receding farther, and being longer, than the one on the right. ■

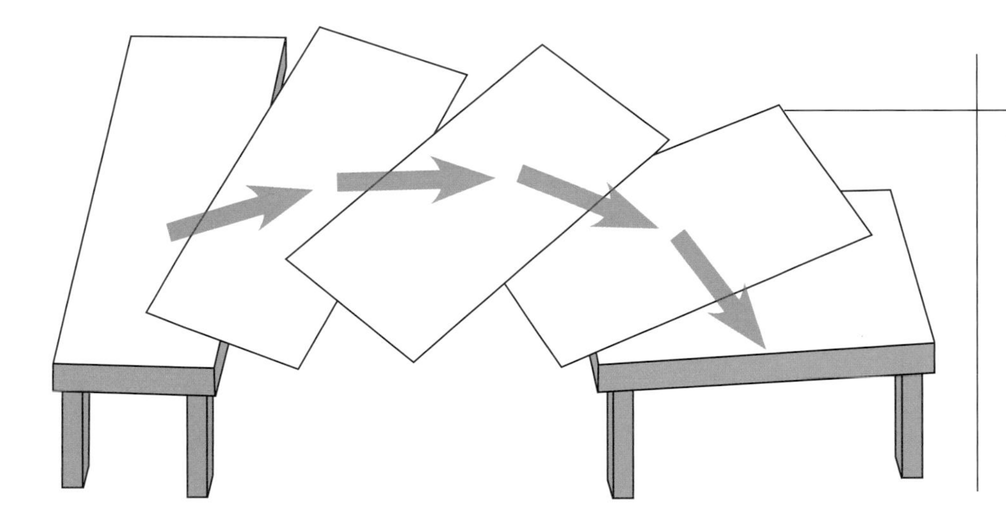

Shepard tables

The left-hand tabletop appears longer and thinner than its mate until it rotates, but the effect is just an illusion. It remains identical to the right-hand tabletop. Hold a ruler to the page to confirm the measurements of each side. Roger N. Shepard first published this illusion in his 1990 book *Mind Sights.*

Perspective illusions

Anybody who has played around with a camera knows that when the three-dimensional world is flattened into a two-dimensional image, optical distortions get introduced. Bringing nearby and faraway objects into close proximity distorts their true sizes and distances and teases the eye of the beholder.

Look closer

Two common misinterpretations are at work in the Shepard Tables: a horizontal/vertical illusion and an illusion of three-dimensional perspective.

Stanford University cognitive scientist and psychologist Roger N. Shepard has created dozens of brain-tickling designs with pen and ink. He began drawing for recreation but found that his images connected to his research on the brain's processing of visual sensations. Shepard delights in finding ways to make the mind "flip" between competing interpretations of visual stimuli or perform what he calls "mental somersaults."

His so-called Shepard tables, featured in the illusion that opened this experiment, demonstrate how the mind attempts to see the world in three dimensions even when confronted by a two-dimensional drawing. It's the same process at work when a tourist snapshot depicts a traveler, enlarged in the foreground, apparently pushing against a tiny Leaning Tower of Pisa in the background as if to shift it upright, or a hiker in Utah's canyon country appearing to hold up a boulder that, in reality, lies at a greater distance from the camera.

Puzzling out the illusion

The horizontal and vertical lines of the Shepard tables exist in nature, but not in equal proportion. The eye is much more likely to see a long horizontal, such as the horizon line or the far edge of a lake. A long vertical is more likely to represent something horizontal on the ground, such as an open road that recedes into the distance, and thus we are more likely to interpret it as longer than a true horizontal.

Fick's T

German physiologist Adolf Eugen Fick (1829–1901) created this illustration as a parlor trick. The height and width of the inverted T are exactly the same, although most people think the vertical line is longer than the horizontal one. The top hat follows the same principle, highlighting the way the eye and brain process each axis differently.

The vertical lines in Shepard's drawing seem to stretch toward the horizon, like a road. But something's "wrong" with that interpretation. If the tops of the tables were truly rectangular, linear perspective would require the far edges to appear slightly shorter than the parallel edges closer to us. Experience with perspective causes the brain to see objects as farther away if they are higher on the horizon or smaller than similar objects of the same size. In this case, the brain registers the fact that the far edges are the wrong size to match the natural world if the tabletops are perfectly rectangular, but it discards this information to accept the best fit for the data: two real tables seen at different angles.

The arrow illusion

Compare the Shepard tables with the Müller-Lyer, or arrow, illusion, named for 19th-century German psychiatrist and sociologist Franz Carl Müller-Lyer. He began his illusion by drawing two parallel lines of the same length. At the ends of one line, he placed two arrowhead shapes with their open ends pointing outward. At the ends of the other line, he placed two arrowhead shapes that were identical except their open ends pointed inward. The line segment with the arrowheads pointing inward and the ends open to the outside looks significantly longer than its mate. The illusion holds true with the line segments in any orientation.

Scientists aren't sure why this illusion exists, but they have theories. One is the limited acuity theory. Observers would expect the line segment with the inward-pointing arrowheads to look longer if the eye's acuity outside the area of greatest focus—the fovea, directed at the line segments themselves—were of a resolution low enough to soften the edges of the images. When a piece of sharpness-softening tracing paper is placed atop the Müller-Lyer illusion, the line segments

Railroading

We've all seen railroad tracks recede in the distance. When we see two lines converging, our brains often read the convergence as distance, even when this misperception distorts the actual image. This illusion makes it difficult for us to judge widths accurately, even in a diagram as simple as the one below.

CASE STUDY

Cézanne's blank spaces

A bowl of fruit. A mountain in France. Two men playing cards. What's so revolutionary about any of those?

The genius of French artist Paul Cézanne (1839–1906) was to see in a new way. Earlier painters had sought a sort of photo-realism, capturing the tiniest details frozen in time, or they tried, like the Impressionists, to fix the transitory qualities of light. Cézanne dug deeper. He realized that light is the beginning, not the end, of vision.

"The eye is not enough," Cézanne said. "One needs to think as well."

Cézanne paintings have bold brush strokes but no sharp boundaries. Colors push against each other, subtly shift and blend, and sometimes disappear entirely, leaving bare patches of canvas in which the viewer has to fill in the blanks. Yet they depict recognizable objects, people, and places. Our brains help resolve the chaos.

According to Jonah Lehrer in his book *Proust Was a Neuroscientist,* Cézanne painted the world as it appears before the mind's interpretations have resolved it. Significantly, Cézanne developed his ideas at about the same time the Gestalt psychologists formulated their theories about vision's being more than just the sum of sensations. Fellow painter Émile Bernard, one of the few to witness Cézanne at work, said the revolutionary artist "only interpreted what he saw, he did not try to copy it. His vision was centered much more in his brain, than in his eye."

As Cézanne got older, he left more and more blank areas on his canvases. Calling the works unfinished, critics scoffed. But Cézanne knew that the paintings supplied everything that a viewer needed. He thought hard to figure out how the brain would interpret empty spots. Thus, when Cézanne kept an area free of paint, he had already determined how the brain would fill it in.

Cornered

When arrows are added to the tops and bottoms of two lines of equal length, our brains are tempted to perceive these differences as three-dimensional cues. The arrows on the outside seem to come toward us, while the arrows on the inside seem to go away from us.

may appear to have slightly different lengths. However, the change is so small that it cannot entirely explain the illusion.

Cornering a theory

Another theory is the corner illusion. People living in developed countries see right angles all the time. The brain interprets three lines converging at a point as the corner of a room. When the eyes look at the Müller-Lyer illusion, the brain draws upon its experiences to interpret the arrows as depth cues.

One way to test the probability that false perspective contributes to illusions is to find people whose brains have never experienced optical cues of distance

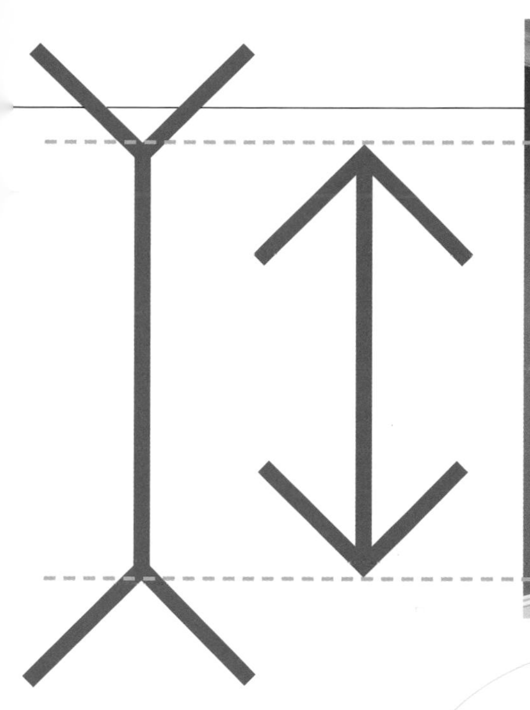

perspective or seen buildings with right angles. Tribal peoples in rain forests have little or no experience seeing faraway objects because they live in small openings in the trees. When taken to open ground for the first time, they see distant objects as small instead of far away. Rural Zulus live in a culture nearly devoid of straight lines. Their huts and fences are round. When tested, rural Zulus have a weaker reaction to the Müller-Lyer illusion than Westerners do. ■

The takeaway

Your brain interprets the images your eyes see. In particular, a lifetime of visual experience stored in your memory strongly influences how you decode the meaning of what you see. You interpret horizontal and vertical lines differently, and you tend to impose perspective when you see two-dimensional representations of the three-dimensional world. When your interpretation of visual sensations doesn't match your understanding of reality, you get tricked by illusions such as the Shepard tables.

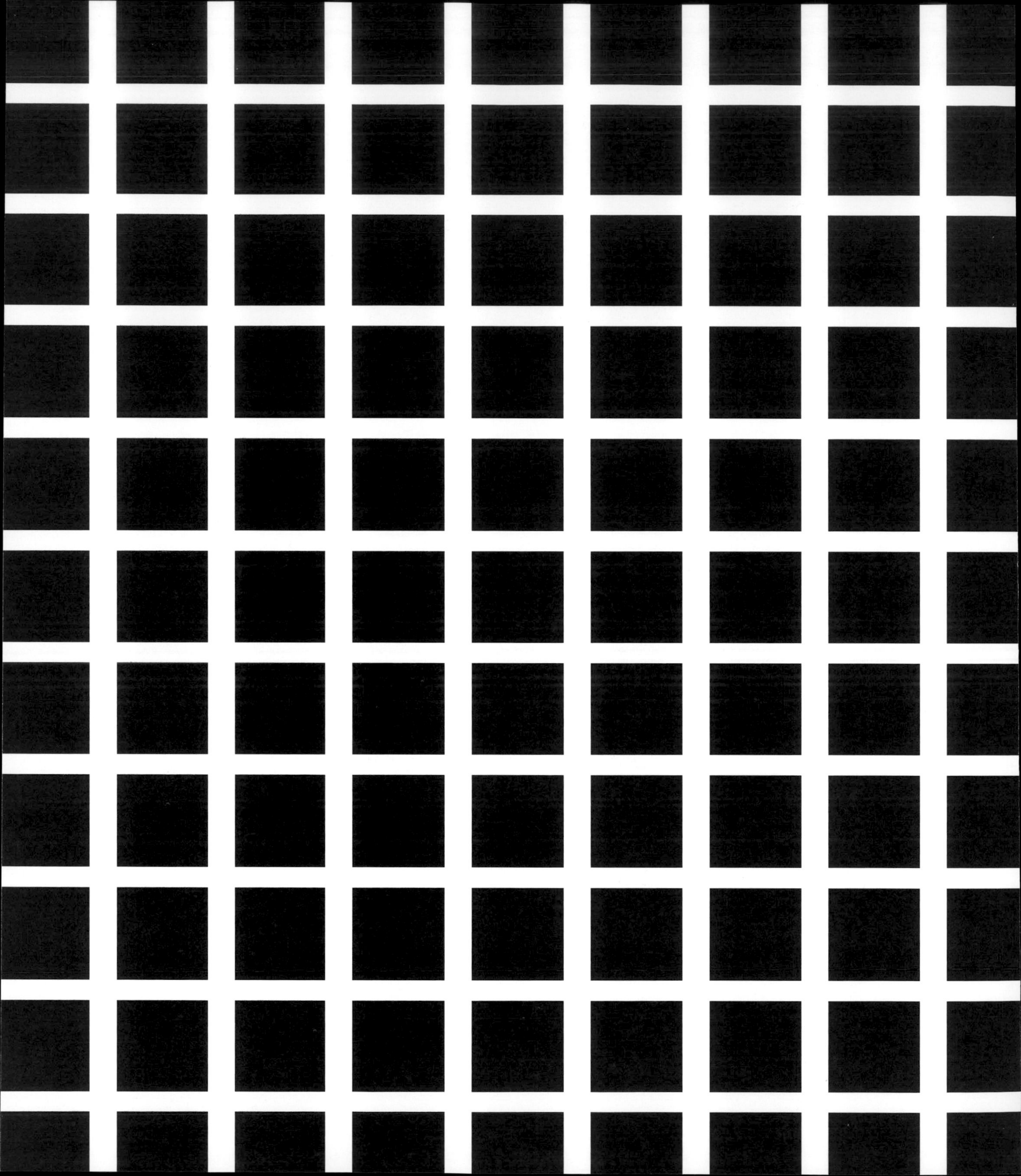

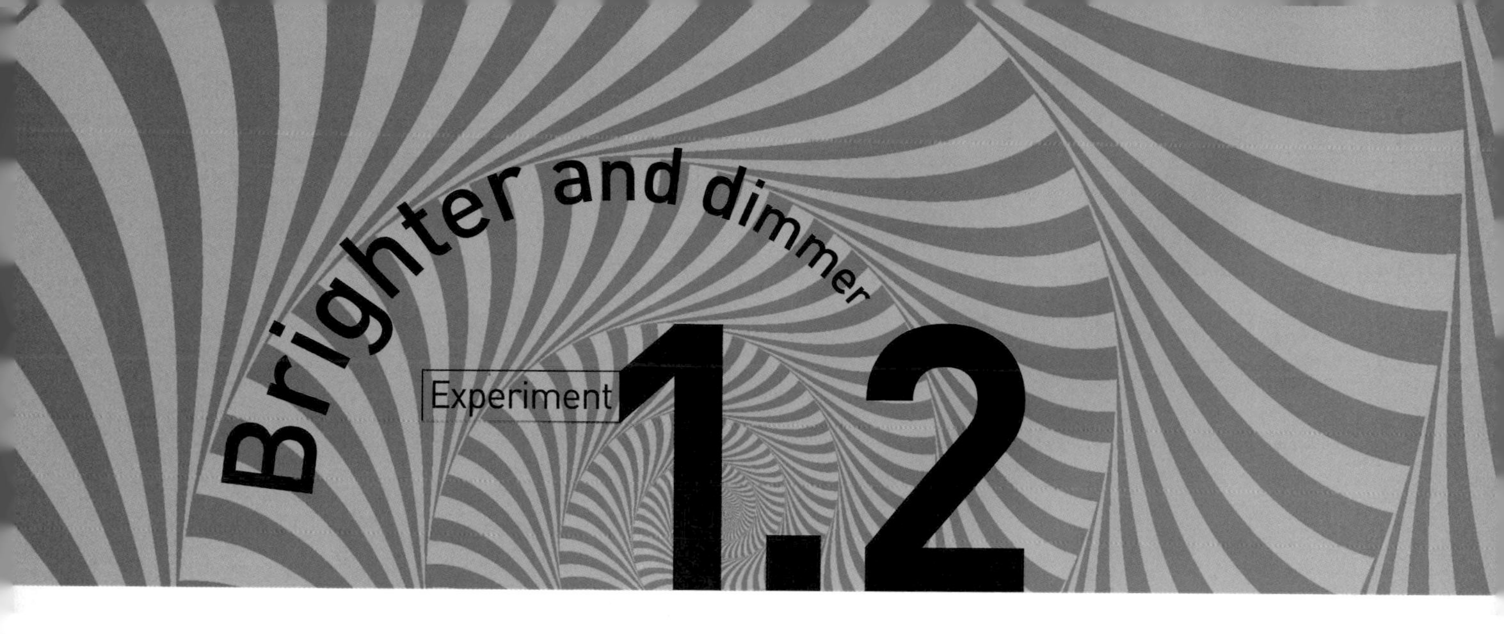

Brighter and dimmer

Experiment 1.2

Let your eyes linger

on the grid of black squares separated by perpendicular white bars.

Focus your attention

directly on a particular black square, or just relax and try to take in the entire figure at once.

When you use your peripheral vision,

do you notice anything unusual about the spaces between the squares or the intersections of the white bars at the squares' corners?

Now turn the page.

What happened

In the regions where the white bars intersect, ghostly gray phantoms appear and then dissolve back into nothingness. These slippery ephemera seem real until you try to fix one in your sights by staring directly at it. Then it disappears.

You can prove the phantoms are not on the page by covering two rows of black squares with white paper, or by using a sensitive light meter to take readings from the intersections and from the region surrounding the grid.

The explanation for the illusion lies in the organization and function of light-detecting neurons in the retina. Some of the retina's most sensitive photoreceptors fire in response to light but lessen their response when nearby photoreceptors fire at the same time. The process in which neurons suppress their neighbors' actions is known as lateral inhibition. Where white and black come together, lateral inhibition causes the white to lose some of its brightness and to appear gray.

The gray disappears when you stare directly at it because the fovea—about the size of this letter *o*—is less sensitive to light than the rest of the retina and therefore is less susceptible to lateral inhibition. So, when you look directly at the ghostly gray shapes at the intersections of the grid, they likely disappear.

Gray versus gray

The two central gray squares reflect the same amount of light. Due to the effect of simultaneous contrast caused by lateral inhibition during retinal processing of the different backgrounds, however, the gray on the light background appears darker.

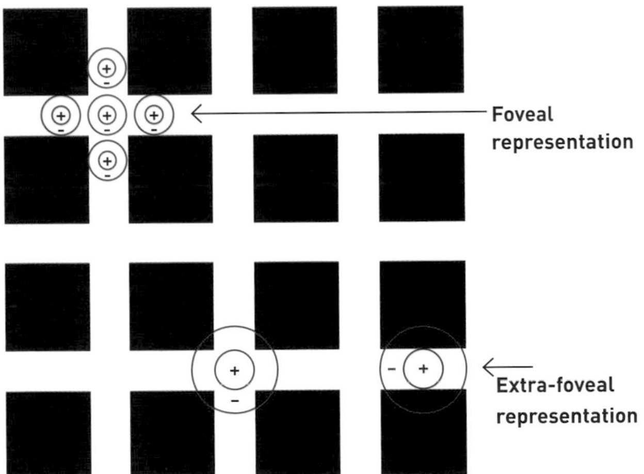

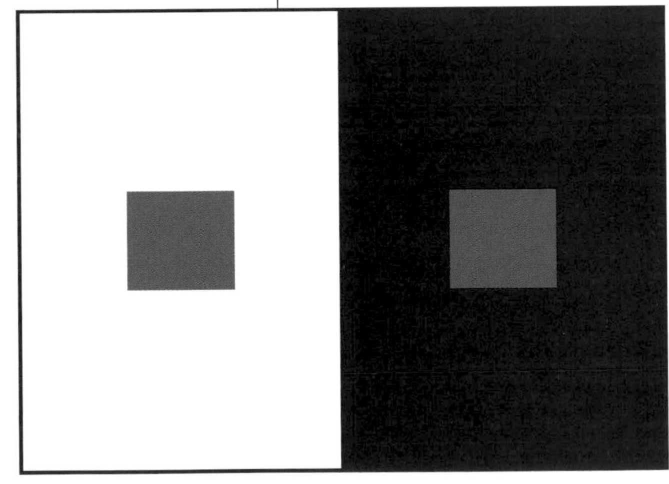

Optical illusion gallery: **http://www.michaelbach.de/ot/**
Broad range of visual phenomena, with supporting scientific explanations.

German physiologist Ludimar Hermann discovered this phenomenon in 1870 while reading a book. The book's author had arranged figures in a matrix on a page. When Hermann stared at the numbers, he saw gray spots at the intersections. He investigated the phenomenon, reported his analysis in scientific literature, and gave his name to the illusion that opened this experiment: the Hermann grid. Others have used the Hermann grid for science and fun. A century after Hermann's discovery, Professor Günter Baumgartner used the grid as an instrument with which to measure the size of the human retina's receptive fields. In 2000, shortly after the disputed American presidential election, a modified version pinged around the world as an email attachment, asking recipients to count transitory dark and light spots as votes for Al Gore or George W. Bush. ■

Look closer

Three investigators—Keffer Hartline, Floyd Ratliff, and Henry Wagner—demonstrated the existence of lateral inhibition in 1954.

Working with the simple eyes of horseshoe crabs, they found that a photoreceptor's electrical output decreases with stimulation of nearby photoreceptors. A nerve bundle called the lateral plexus connects the photoreceptors and allows them to influence each other.

The human eye is much more complicated, but the principle is the same. The human retina contains five kinds of vision neurons: receptors (rods and cones), horizontal cells, bipolar cells, amacrine cells, and ganglion cells. When light hits rods and cones, they send signals that travel to the bipolar cells and then to the ganglion cells, which collect pulses for relay to the optic nerve. Horizontal and amacrine cells also interact with photoreceptors. This arrangement lets one neuron communicate with many others. Convergence among rods heightens sensitivity to light. Convergence among cones heightens the ability to see fine details. In particular, the phenomenon enhances the perception of edges.

Sometimes information gets added during neural convergence. Sometimes, strangely, the brain seems to toss it out. Why this is so springs in part from the evolutionary advantages of seeing edges, such as the lip of a chasm, as sharp and clear. If a light-colored object, such as a rock, lies next to a dark space, such as a shadowy canyon, lateral inhibition makes the canyon's darkness seem darker and the rock seem brighter. The rock's edge stands out, and you don't tumble into space.

Living on the edge

Edges between dark and light fill the Hermann grid. Psychologist E. Bruce Goldstein provides an explanation of the ghostly grays at the grid's intersections. Consider a point at one

CASE STUDY

Motion blindness

A rare brain condition called akinetopsia (or motion agnosia) results from damage to the neural networks associated with perceiving motion. These networks lie near the junction of the temporal, parietal, and occipital lobes. Detecting motion is crucial. Viewed through an evolutionary lens, it provides survival benefits such as perceiving the approach of something dangerous (predator) or beneficial (prey).

In 1983, neuroscientists published a paper describing a middle-aged woman, referred to in the literature as L.M., who developed a vascular lesion in a cerebral region near the striate cortex. She lost the ability to recognize when objects were in motion, even though she could recognize them when they stood still. L.M. told researchers of her difficulty pouring tea or coffee into a cup. Because her brain could not recognize the motion of the liquid rising against the cup's interior sides—the liquid appeared frozen to her—she did not know when to stop pouring. L.M. also had difficulty following conversations. As she watched the faces of people speaking, the words seemed to emerge from immobile masks.

The most disconcerting effect of L.M.'s condition occurred when doctors, nurses, and visitors stepped into her room. Normally, a patient would see people approach and walk toward the bed. L.M. saw people blink in and out of existence around her. The phenomenon became dangerous when she looked at traffic, as when trying to cross a street. She had no way to judge the approach of faraway cars, as they winked out and suddenly appeared nearby.

The weird world that greeted L.M. whenever she stepped outside affected her so deeply that she rarely left the comfort of her room.

Bright and dark

Lateral inhibition pools sensations of dim lights so the human eye can detect them. By sharpening edges between regions of darkness and light, it helps you distinguish between sparks of fireworks and the blackness of surrounding sky.

intersection, he says. Let's call it point A. Assume the light falling on A produces an electrical signal of 100 units in a bipolar cell. Now consider each of four points—B, C, D, and E—to the left, right, above, and below A. Each is positioned not at an intersection, but between two black squares. Each of the four points receives the same 100 units of response. If we assume that the bipolar cell for each of these four points produces 10 units of lateral inhibition, then point A's bipolar cell's response will be decreased by 40 (4 x 10) units, yielding a signal of 60. The bipolar cells for B, C, D, and E also are subject to lateral inhibition, but less so than A. Bright areas above and below, or on

either side, strongly inhibit the bipolar cell's response for each point, and the adjacent dark areas add a weaker inhibition. The signal for B, C, D, and E will be reduced by 20 (2 x 10) units from the two neighboring bright areas, and a little bit more from the dark areas, for a total just below 80. Greater inhibition for A results in the perception that the intersections are darker than nearby white spaces. ■

The takeaway

Neurons in the retina influence each other through lateral inhibition. This sharing of information lets rod cells pool faint signals and detect a weak light on a dark night. It allows cone cells to compile a dense mosaic of sharp, colorful detail in daytime. It also helps the brain detect the outlines of objects more readily. A minor side effect of these abilities is the existence of optical illusions such as the Hermann grid.

Afterimages

Experiment 1.3

Stare

at the star on the oddly colored photograph of a Spanish castle.
Keep looking, without moving your eyes, and count off 60 seconds in your mind.
You can blink your eyes, but try not to look away.

Now shift

your gaze to the accompanying black-and-white photo of the same castle.

What happens

to your perception of the grayscale photograph of the castle when you transfer your attention to it after viewing the first photograph?

Now turn the page.

What happened

If you have normal color vision, the black-and-white photograph probably appeared to be in natural color, with a blue sky and green grass. After a few seconds, it returned to black and white.

Neurons that detect color, like other sensory neurons, undergo adaptation after prolonged exposure. They get tired of sending the same information and reduce the intensity of signals they send to other neurons. When you shift your gaze, your tired ganglion cells react more strongly to colors that form either-or relationships with colors suppressed by adaptation. These either-or relationships are known as the red-green and blue-yellow opponent pathways. A ganglion processes only one of the pairs at one time. Thus, it can signal red or green, but not both simultaneously. If adaptation weakens the ganglion's red signals, it processes green more strongly when given the chance. Your eyes briefly grow more sensitive to light stimulating the green end of the red-green pathway, even if that light reflects off something not entirely green.

Using photographic software and knowledge of adaptation, graphic artist John Sadowski created the Spanish castle illusion. Sadowski posted a version of this illusion on his personal website and invited comments. Several readers claimed the website substituted a different photo a few seconds after they looked at the original, black-and-white one. Your duplication of the illusion with a printed page disproves that misconception. ■

Bits of color

Color registers on the retina as bits of energy that activate cones particularly sensitive to red, green, or blue. These bits form a whole, in much the same way colored dots in the close-up of a printed page or TV image merge in a way the brain experiences as all colors.

Look closer

Color vision is extremely complicated. It begins with the stimulation of cone cells in the retina and gets modified by specialized information-processing neurons in the retina and the visual cortex, as well as other parts of the brain.

Two key theories, the trichromatic theory and opponent-color theory, explain much of what scientists know about color vision. These two theories complement each other, and researchers have found solid evidence to support both. The first describes how sensation of all colors that humans see begins with three kinds of cone cells in the retina. The second begins to explain how neural mechanisms receive impulses from the cone cells and modify them into complex colors.

The trichromatic theory

Two 19th-century scientists, Thomas Young and Hermann von Helmholtz, proposed using experiments with light to explain color vision. They found that when experimenting on people of normal vision, their adjustments of three separate wavelengths of light could duplicate the perception of any color created separately by a single wavelength. For example, mixing bits of red, blue, and green light might create a very particular shade of green. According to the theory, proposed by Young in 1802 and later refined and championed by von Helmholtz, color vision depends on three sets of color receptors (hence the word *trichromatic*) in

Light's spectrum

Facets of a diamond break ordinary white light into the colored bands of the rainbow. Color doesn't physically exist in an object. Instead, part of the bandwidth of light reflects off the object and stimulates some cone cells strongly, but others less or not at all. Your brain interprets the result as color.

the retina. Each receptor reacts to light entering the eye in a particular range of frequencies, or wavelengths.

Physiologists found hard evidence to back up Young's and von Helmholtz's theories in the 1950s and 1960s. They discovered that light-sensitive pigments in three kinds of human cone cells react selectively to light. Some retinal pigments react most strongly to long wavelengths, some to medium wavelengths, and some to short wavelengths in the visible spectrum. In scientific shorthand, these pigments are found in the L (long), M (medium), and S (short) cones. These cones sometimes are assigned color names—with L as red, M as green, and S as blue—but that's a bit of an oversimplification. All cones react strongly to particular colors and more weakly to others. All visible colors are formed by mixtures of strong and weak reactions among the three cone types in varying intensities, with strong reactions reflected in the frequency of a cone's firing rate. White light, as Sir Isaac Newton's experiments with prisms

CASE STUDY

Seeing new colors

Humans have three kinds of cones. Birds have four, which extends their vision into the ultraviolet spectrum. What would you see if you could see like a bird? Ornithologist Geoffrey E. Hill says people frequently ask him just that question. His response: "We can no more appreciate what such color looks like in the brain of a bird than we can appreciate how a world perceived by ecolocation appears in the brain of a bat."

And yet the question remains. Jay and Maureen Neitz, professors at the Medical College of Wisconsin, have worked on adding a third cone to the eyes of monkeys that have only two. They created a virus that contains the DNA code for the gene that would provide monkeys with a third photoreceptive pigment and injected the virus into the monkeys' eyes. If the monkeys' retinas incorporate the virus DNA, they should be able to detect new wavelengths of light. And if their brains' visual networks can process the new information, the monkeys would see colors they have never seen before.

That made the Neitzes wonder: If a virus could add a third cone pigment to a monkey, why not a fourth to a human? The result would create tetrachromatic vision. If that indeed becomes possible, humans would be able to see in two additional colors. If the new cones extended vision into the infrared portion of the spectrum, humans would gain some ability to see in the dark. It's uncertain whether a human brain would accept the new input. But Jay Neitz is confident that given the brain's plasticity, it could.

> Illusions occur when our brains attempt to perceive the future, and those perceptions don't match reality. Mark Changizi

demonstrated, contains all colors, so it stimulates all three kinds of cones very strongly.

Color blindness

It is interesting to note that the pigments in the L and M cones are genetically encoded on the X chromosome, one of the two that determine a person's sex. Women have two X chromosomes and pass one to their children. Men have one X chromosome and one Y chromosome, and also pass one of the two to their children. The child who gets an X from the mother and an X from the father is a girl; the combination of X and Y results in a boy. This explains why the vast majority of people with color blindness are males. The gene for normal cone cells is dominant, so anyone with a color-blind X chromosome and a normal X chromosome would likely see a full range of colors. However, a man with a color-blind X has only a Y chromosome, with no genetic codes for color vision, and thus is color-blind.

Most often, people who are color-blind lack just one type of cone. Such people are called dichromats. The most common dichromatic condition creates an inability to distinguish between red and green. People with this type of color blindness see what others would describe as a muddy mix of blues, yellows, and grays. A much rarer kind of

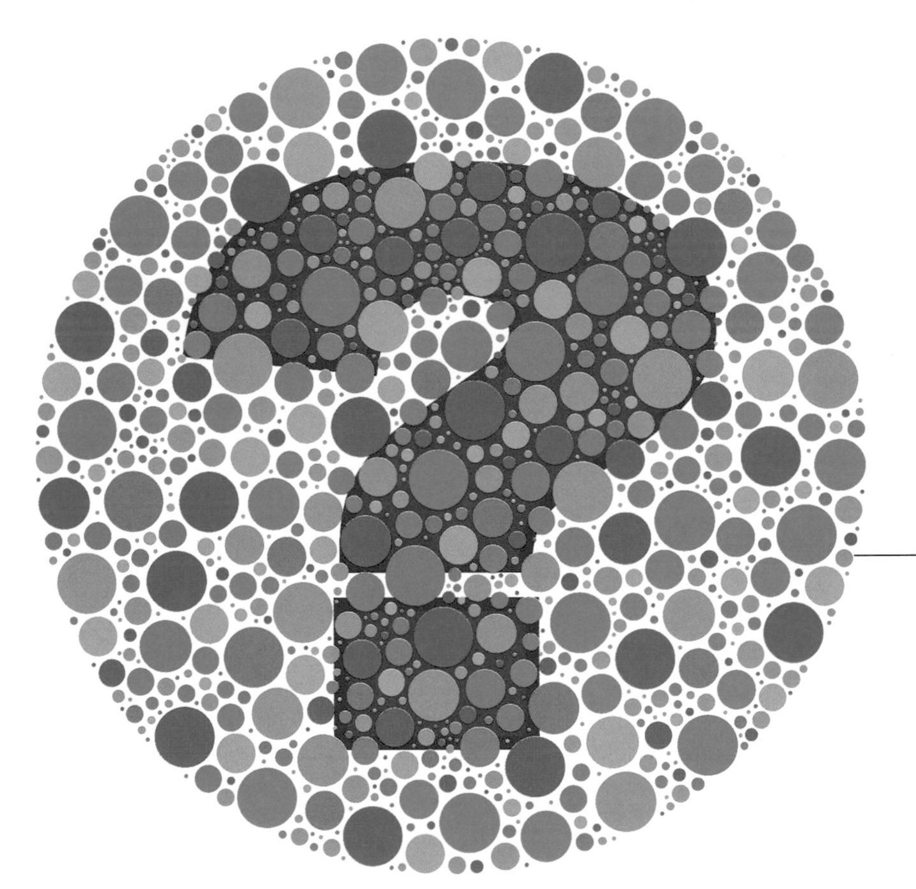

Healthy cones

Color blindness occurs when some or all of the three types of cone cells function weakly or not at all, or perhaps are absent. Red-green dichromatism impedes the ability to distinguish these two colors. A simple test (left) can diagnose it.

Opponent colors

Ewald Hering, who posited all colors as a mix of red, yellow, green, and blue, created opponent color diagrams (right). Hering said we see one-half of a red-green or blue-yellow pair at a time.

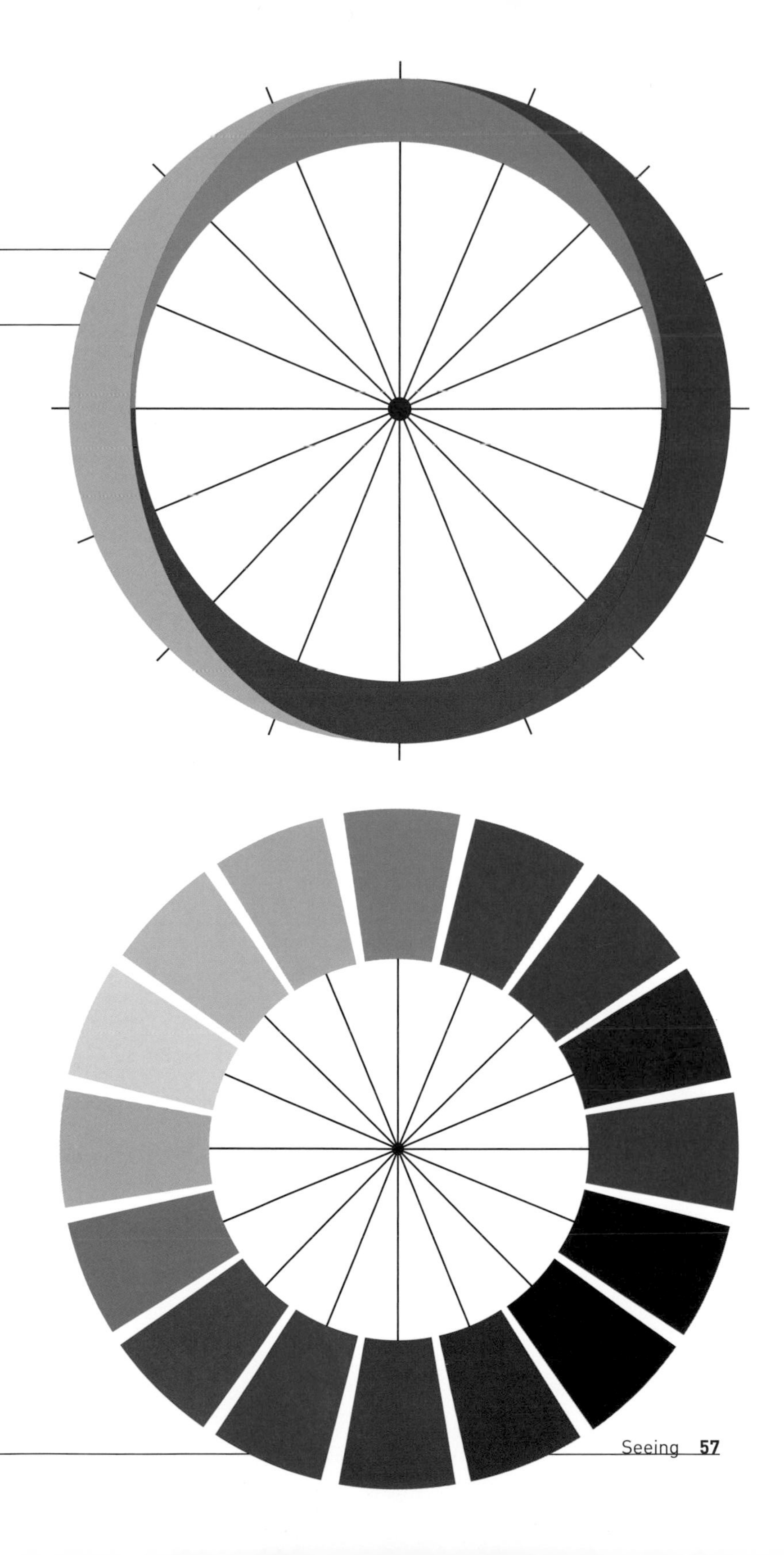

dichromatism, blue-yellow, is not related to the sex chromosome. In each case, however, a color-blind person lacks function in one or more sets of cone cells, lacks cone cells completely, or has problems processing color further along in the brain's visual system.

The opponent-color theory

Ewald Hering, another German physiologist, expanded on the work of von Helmholtz, his contemporary. Hering showed a variety of colors and shapes to his test subjects and asked them to describe what they saw. One curious observation was that prolonged viewing of a green field created a red afterimage, as you have demonstrated with the Spanish castle illusion.

Blue flower, red thorns! Blue flower, red thorns! Oh, this would

Hering's work led to his 1902 theory of the opponent process of color vision, which holds that the brain's perception of colors depends upon physiological responses to opposing blue-yellow and red-green pairs.

Scientists have found physical evidence for Hering's opponent-process theory in the retina and the lateral geniculate nucleus. There lie neurons that express opposite reactions to paired wavelengths of light. One such set of neurons is called B+Y- because blue light excites its firing response and yellow light dampens it. Opposite neurons, called B-Y+, also exist. In addition, there are red-green pairs, represented as R+G- and R-G+. The pairing of these opposites explains why you can see colors that are reddish and bluish, such as violet. When that happens, both blue and red sensors activate at the same time. As green and red cannot simultaneously respond, you cannot see a color that appears reddish-green.

BREAKOUT

Types of color blindness

Protanomaly: A color deficiency also known as red weakness. Red and its component in other colors is seen weakly, both in saturation and in brightness. Violet may appear as blue, and orange and yellow appear shifted toward green.

Deuteranomaly: A color deficiency also known as green weakness. Colors maintain their brightness but appear shifted toward red, causing poor discrimination among red, orange, yellow, and green.

Dichromacy: No perceptible differences among red, orange, yellow, and green.

Protanopia: Reduced brightness of red, orange, and yellow. Causes confusion of reds with grays and blacks, and reduces or removes the ability to distinguish among hues that contain red.

Deuteranopia: Similar to protanopia, but without reduced brightness. Red, orange, yellow, and green appear much the same color.

Complete achromatopsia: Allows a person to see only white, black, and shades of gray due to completely nonfunctional cones.

Incomplete achromatopsia: Like the complete condition, except weak function in some cones provides a hint of color vision.

Red and green marbles

Neuroscience for Kids, an online curriculum for schoolchildren developed by the University of Washington, offers a simple and delightful analogy for the opponent-color theory. It suggests we think of axons in the optic nerve as tubes or channels. Some carry only streams of red or green marbles, but never both colors at the same time. Others carry only blue or yellow marbles, but never both at the same time. (A third tube carries marbles that are bright or dark, signifying the

Center-surround

Ganglion receptive fields, represented in diagrams (right), illustrate opponent color pairs. Some react to stimulation at the center, others at the edges. If you stare at the plus sign amid the colored squares, then look at the white squares, you should see opposite colors.

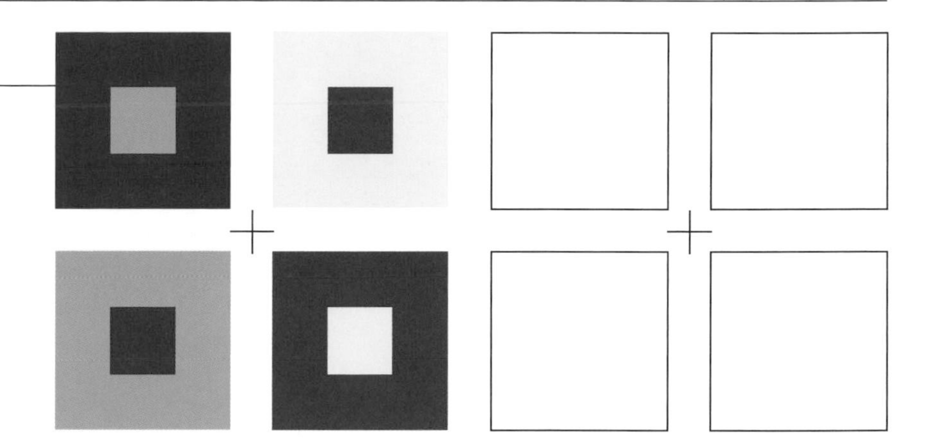

luminance of an object, which is a function of the firing rate of photoreceptors in the retina.)

When marbles emerge from the tubes at the visual center at the back of the brain, red ones push buttons that make the brain see red, blue ones push blue buttons, and so on. Red and green marbles cannot pass through the tube at the same time, so red-green neurons cannot simultaneously react to red and green. The same is true of blue and yellow. However, the Washington educators note, the yellow marble is an oversimplification. Ganglion cells working with neighboring cells in the retina create the color yellow from a combination of red-firing neurons and green-firing neurons situated side by side—the L and M cones. The brain blends and modifies information in the red-green, blue-yellow, and bright-dark neural pathways to create colors of all hues and intensities. Much of the processing of the original L, M, and S cone cells occurs in the V1, V2, and V4 regions of the visual cortex. ■

*The takeaway

Neurons grow tired after long exposure to monotonous sensations. As a result, they reduce their response, a phenomenon known as adaptation. If you stare at something and look away, adaptation creates an afterimage substituting blue for yellow and red for green. This afterimage serves as evidence supporting the opponent-color theory of vision. A related theory suggests the brain creates colors by manipulating information from three types of cone cells.

Study

the artwork depicting two sets of crossed bars above a white-and-black checkerboard.

Choose

a name for the color where the bars intersect in the left-hand illustration, which lies under a yellow mask. Now name the color in the similar intersection in the right-hand illustration. Also examine the four angles where the bars cross.

Decide

whether you would call the angles acute and obtuse, or perpendicular.

Now turn the page.

What happened

Did you call the color in the left illustration blue and the one in the right yellow? Congratulations. You probably have a fairly normal interpretation of the visible spectrum of light.

Your retinas and your brain react the "right" way: They assign the most commonly selected colors to the light reflected by the two illustrations.

You're also wrong. The two colors are exactly the same shade of gray. Likewise, the two sets of colored sticks intersect at 90-degree angles.

Even now, reading that the colors are identical may elicit skepticism. But you can prove this to yourself. Get a few index cards or a pad of gummed memo papers, the kind used for sticking notes on desks and papers, and place them around the central elements of the two crosses to completely isolate them from the surrounding colors. The blue becomes gray. So does the yellow. To check the angles of the intersecting banded sticks, use the edge of an index card or a piece of paper. Each is a right angle.

Your brain judges color through its experience of color. In nature colors don't exist in isolation, a fact that becomes important when the paint you liked in the hardware store doesn't look right on your bedroom walls. In making interpretations, your brain seeks cues such as the quality of light and the context of adjacent colors. Manipulation of context and light can alter perception so that a particular color appears radically different, even when presented near its twin. Similarly, cues suggesting three-dimensional perspective influence your perception of lines and angles. ■

Color context

R. Beau Lotto's illusions explore how perceptions of brightness and color spring from context. The same stimulus can appear light or dark. Likewise, shading and color can suggest obtuse or acute angles.

! Look closer

This illusion is the work of R. Beau Lotto, a neuroscientist who mixes optics and art to create stunning and informative visual displays in galleries and public spaces. He knows the illusion's power arises from the brain, not the eye.

"Context is everything, because our brains have evolved to constantly re-define normality," Lotto told the *Times* of London. "What we see is defined by our experiences of the past, but also by what the human race has experienced through its history. The structure of the brain is a reflection of that history."

Throughout history, painters have known that colors look different when they appear next to other colors on their canvases. One method they use to assess the qualities of color is to turn their paintings upside down, as this removes some of the context of object recognition that might interfere with judgment. Qualities of light also alter perception of color. Thus, the only true way to know how a painting will look

Beau Lotto's Lottolab: **http://www.lottolab.org/**
Explores phenomena of color and brightness.

> Alas, our brains haven't evolved much over the past 50,000 years, and we're stuck with ancient hardware. Jonah Lehrer

on the wall above your fireplace is to hang it there and observe it under various conditions of natural and artificial light. Warm incandescent bulbs and cool fluorescent tubes change the appearance of a painting, as do various intensities.

Nature or nurture?

Scientists once suggested that innate wiring of neural pathways leads to the perception of color differences. Lotto points instead to the brain's ability to accumulate experiences about the world and then to use them to make its best guess about what it sees. Through trial and error, some experiences occurred over eons to create brain structures that give us evolutionary advantages. Thus, newborn children have no experience in the world outside the womb, yet they react reflexively to faces. Other experiences

Contrasts

Background affects color. For impact, put secondary colors against constituent primaries: Orange looks more red on yellow.

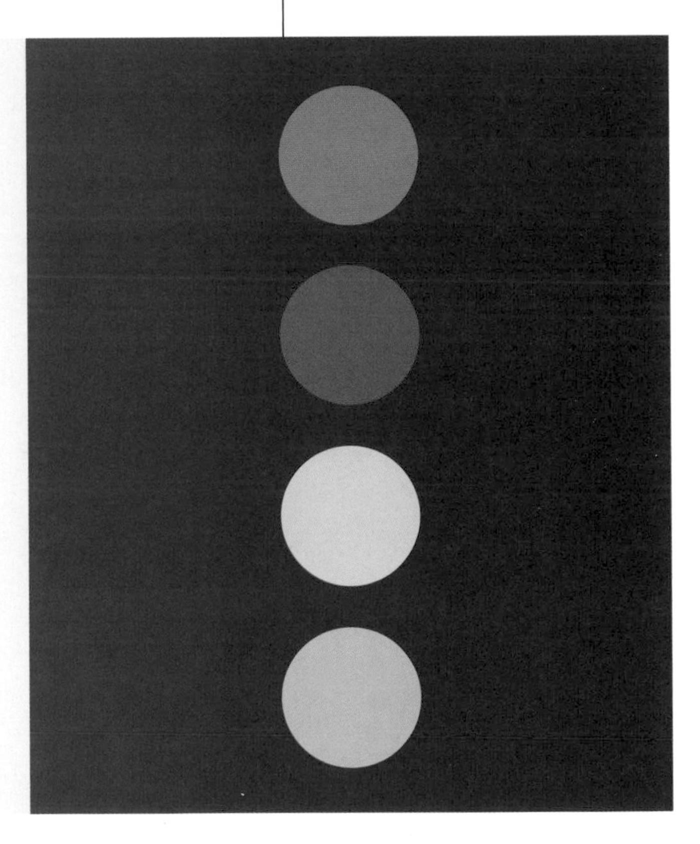

accrue as babies age. Their brains store memories of visual stimuli. They match new stimuli with stored knowledge to arrive at probabilistic conclusions about what they see.

This process occurs, Lotto says, because of the ambiguity of vision. The eyes and the brain have no direct contact with the outside world. They can only render three-dimensional objects into two dimensions on the retina. However, an infinite number of three-dimensional shapes could cause a two-dimensional rendering. When you see an oval, for example, it could represent an oval drawn on a piece of paper or a circle seen from a sharp angle. A line segment could be a stick, a box edge, or even the rim of a circular object, such as a tire or a quarter, seen from an unusual angle.

One shape, many objects

One of the most compelling examples of this ambiguity is the Ames room. American

CASE STUDY

The *maksun* of Pingelap

Blame it on the storm of 1775. That year, Typhoon Lengkieki crashed across Micronesia, a collection of Pacific atolls near the Equator. The storm and its saltwater surge destroyed the taro crop and the banana, breadfruit, and coconut trees of Pingelap, a tiny and isolated island. Flooding and starvation killed nearly all of the island's 1,000 inhabitants. Only about 20 remained alive several weeks later.

Eking out a living on their devastated atoll, the survivors caught fish and replanted their orchards and taro. The island revived, but the typhoon played a long-lasting trick on the people. Their gene pool had shrunk to only a handful of fertile adults. In such situations, any rare genetic traits may spread.

One did. The gene for total color blindness hid dormant in the DNA of the island's ruler, Nahnmwarki Mwanenised. But as he and the other survivors bred, and their children intermarried and bred, the gene found expression far beyond its incidence rate in the outside world. Achromatopsia—the condition of having nonfunctional or nonexistent cone cells, and thus being able to see only blacks, whites, and grays—exists in about 1 in every 30,000 to 40,000 humans. On Pingelap today, it strikes 1 in 12.

The first color-blind children on Pingelap appeared in the 1820s. At age two or three months, they turned their faces from the sun and squinted to keep from being blinded by light bleaching their sensitive rod cells. As they grew, it became apparent that they could not see colors and lacked the sharp vision provided by cone cells. Many could not distinguish letters and never learned to read.

The people of Pingelap called their color blindness *maksun*, meaning "not-see." Those with maksun have proved to be skilled at night fishing, as they are particularly adept at seeing the faint gray flashes of fish swimming in the dark.

When an artist employs geometrical perspective he does not

psychologist Adelbert Ames used his background as a painter to create an elaborate trick on the brain: a distorted room that looks normal when viewed from front and center. The back wall slants away instead of lying perpendicular to the viewer's line of sight, but Ames compensated for this by using perspective cues to make the room appear normal. A person at the most distant corner of the sloping wall appears tiny, with plenty of space overhead; the same person standing in the nearer corner crowds the ceiling like a giant. A child on one side dwarfs an adult on the other because both appear equidistant from the viewer. Perspective plus one particular viewing angle make the room appear rectangular and normal.

Lotto says that the brain uses its experience to assign the best decision about an object's identity—including its color and shape—because that technique has proved useful for the survival of the species.

In the case of Lotto's colorful illusion of the crossed sticks,

draw what he sees—he represents his retinal image. Richard L. Gregory

the brain calls upon its massive data bank of experiences to assign colors and angles to the visual stimuli. It doesn't do this with any deliberation. Rather, the visual cortex reacts reflexively; it automatically chooses the most likely interpretation of the sensations reaching the retinas.

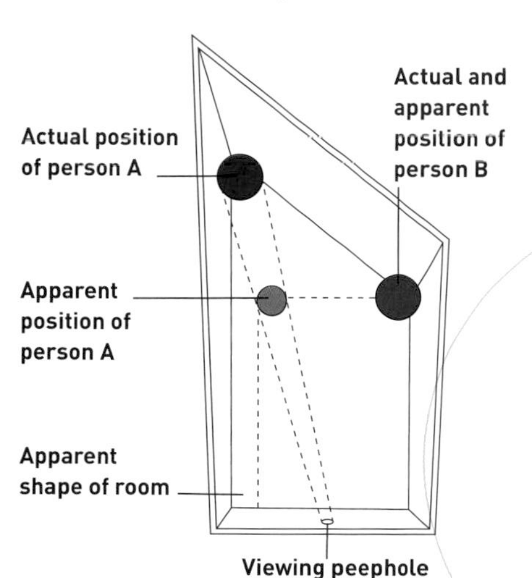

Ames room

Psychologist Adelbert Ames designed a room (above) that appears normal from one vantage point but is in fact distorted. Viewers accustomed to rectangular rooms assume the objects inside it are somehow odd, rather than the room itself.

Within the illustration, clues such as the checkerboard provide perspective, making the likelihood of an acute angle at the top of the left-hand illustration more likely. Color references, in the context of the adjacent colors and the yellow and blue rectangular masks, suggest which hues the brain should assign to the bands at the sticks' intersections.

The fact that the visual cortex does this unconsciously is what makes it so hard for the conscious mind to ignore it. Your willpower cannot override eons of evolution, capped by your own experience. No matter how much your frontal lobes tell you that the colors you see are the same shade of gray, your visual cortex insists they are blue and yellow. ■

*The takeaway

Perception of color, line, and shape is the unconscious work of your brain interpreting what it sees by applying cues of context, light, perspective, and other information compiled through evolution and experience. Babies have to experience the world—and make mistakes along the way—before their brains mature enough to make "correct" interpretations, which are the ones most likely to provide useful feedback. Everyday errors of color interpretation create illusions . . . and complicate home decorating.

Richard Gregory's home page: **http://www.richardgregory.org/**
Vision experiments and videos by the late British psychology professor.

Stare

at the colored gratings on this page for two minutes.
You can let your eyes wander over the vertical green and black stripes
and the horizontal pink and black stripes.

If you like,

make a phone call or listen to music while you look.

Your concentration

isn't required—only your continued visual attention.

Now, turn the page.

What happened

The white bars of the gratings have taken on the colors associated with the original gratings of the opposite orientation. In place of vertical green bars, you likely are looking at vertical pink stripes. And the horizontal white bars of the other grating now appear green.

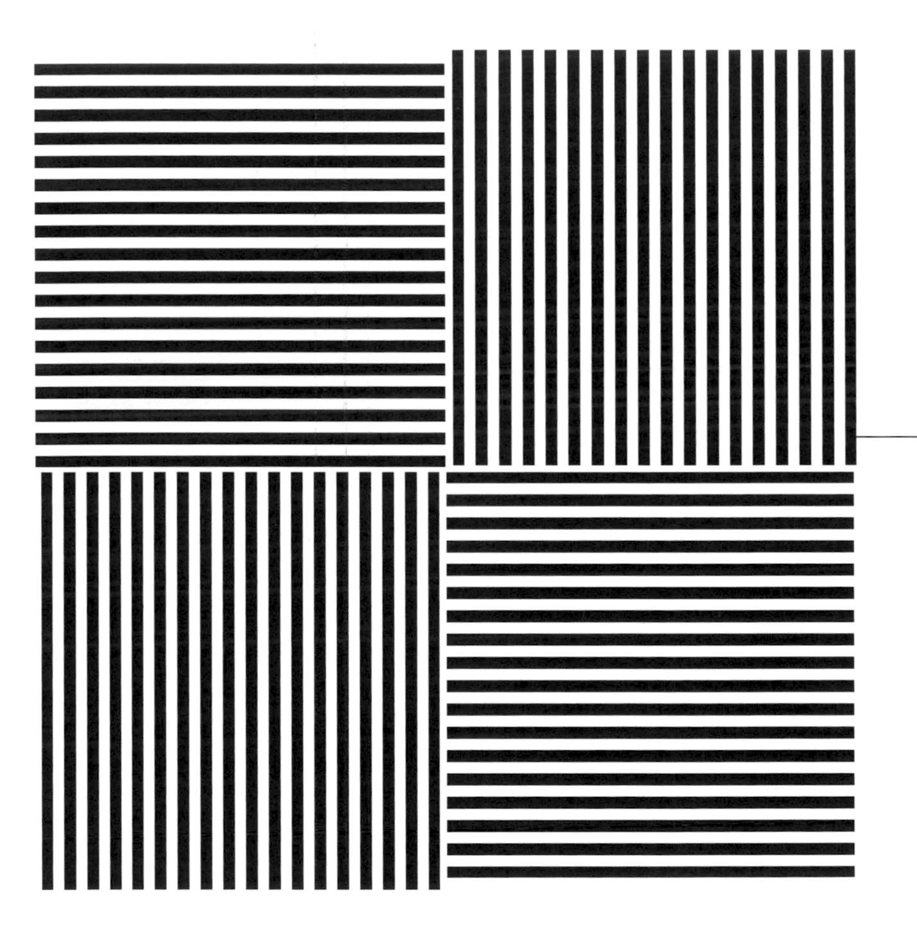

The effect works with other colors, too, and especially well with complementary pairs. For example, blue and orange trade places when paired with the same black and white gratings.

Now comes the weird part. The colors you see as an afterimage depend entirely upon the orientation of the gratings, in a still unexplained phenomenon known as the McCollough effect.

If you try to rotate the page, you see the afterimage colors of the gratings only with their original horizontal or vertical

Contingent color

If you look at the gratings at left without the proper preceding stimulus, they will appear merely black and white. But if you look at them after staring at the image on the previous page, you'll see colors that exist only when paired with vertical or horizontal lines. This phenomenon is known as contingent color. One interesting quality of the contingent effect is that it can last for a long time.

arrangements. In other words, you can't make the pinkish afterimage remain in place on a particular square when you rotate it; all color disappears as the grating turns through 45 degrees, and the other color takes its place as the bars slide perpendicular to their original orientation. Neuroscientists call this a contingent (dependent upon) color aftereffect.

And now, the even weirder part. Test subjects have discovered that hours or days after staring at the colored gratings of the experiment, they perceive colors in black and white gratings. Drinking coffee or taking psychoactive drugs can alter the duration, so some effect likely rests in the neurotransmitters of the visual cortex. And one scientific paper asserted that the afterimage effects appear stronger in extroverts than in introverts. This suggests the illusion could serve as a quick personality test. ■

! Look closer

Psychologist James J. Gibson got the first hint of the existence of contingent color aftereffects in 1929 when he had test subjects wear prisms over their eyes. The glasses shifted the subjects' visual fields about 15 degrees to the right.

Gibson wondered how their brains would adapt to the distortions. He devoted little attention to an unexpected result. In a mere footnote, Gibson wrote that one of his test subjects saw colored bands at the edges of his visual field while wearing the glasses, and the bands reversed position and lingered for hours after the glasses were removed.

Somehow, colors and vertical lines had become linked in the visual cortex.

The next piece of the puzzle appeared in the 1950s. Two American physiologists, David Hubel and Torsten Wiesel, attempted to induce individual neurons in a cat's visual cortex to fire in response to simple shapes projected onto a screen. Hubel and Wiesel had little luck until they pushed a glass slide into the projector, and the cat's glimpse of the slide's straight-line edge caused a particular neuron to fire. They had discovered neurons that respond to lines of light at particular orientations. Some of these so-called edge-detector cells fire only in response to vertical lines, some to horizontal lines, and some to diagonals in between. Hubel and Wiesel's work suggested that the brain constructs the shapes of objects through a complex combination of many neurons reacting in patterns linked directly to sections of an object's outline. By putting the two pieces together—the colored visual aftereffects and the knowledge that some neurons fire in response to specific linear

CASE STUDY

The curious case of Dr. P

The brain and the retinas work together to assign meaning to visual sensations. Light stimulating the retinas sets in motion a chain of events in which millions of electric impulses reach the cerebral cortex. Some signals take a swift route to the prefrontal cortex and give a quick but blurry image to the portion of the brain associated with conscious thought. Some signals get diverted to the visual cortex, which analyzes them for qualities such as line, shape, and movement. The visual cortex then forwards its analysis to the prefrontal cortex, where it arrives about 50 milliseconds after the initial, direct impulses. The prefrontal cortex matches the fast data to its stored experience and creates an interpretation that incorporates the slower data from the visual cortex. In short, the prefrontal cortex tells the rest of the brain what the eyes see. This is why you sometimes look at shapes in a cluttered, empty, darkened room and see people, animals, and even monsters.

What happens when something prevents the brain's higher functions from interpreting visual data? A patient known as Dr. P provided the answer for neurologist Oliver Sacks. Lesions blocked Dr. P's brain from sending nearly any information that made sense of the world. As a result, his brain registered only raw retinal data. It could not interpret the data in any significant way. Dr. P could recognize bits of things such as color and shape, but his brain could not resolve them into objects. When Dr. P rose to leave Dr. Sacks's office, he searched for his hat. According to Sacks, "He reached out his hand, and took hold of his wife's head, tried to lift it off, to put it on. He had apparently mistaken his wife for a hat!"

stimuli—it appeared likely that the contingent colors of Gibson's research might be linked to edge detectors in the cortex.

Long-lasting effects

Psychologist Celeste McCollough made the connection. She had her students at Oberlin College study paired gratings—one with vertical orange bars and the other with horizontal blue bars. She reported her findings in a 1965 article in *Science* magazine: The students saw bluish and yellowish afterimages when exposed to black-and-white gratings. Aftereffects also appear in response to vertical and horizontal lines encountered by chance, such as lines in a book or letters on a poster. Today this lingering color phenomenon is called the McCollough effect.

Many scientists have tried to explain the long-lasting afterimages. One theory holds that the brain has an error-correcting device that compensates for perceptions that do not occur, or occur infrequently, in the natural world. Pink and green stripes certainly don't lurk around every

McCollough decay

The McCollough effect, induced by prolonged exposure to colored gratings (1), decays at variable rates depending on subsequent stimuli. Specific black and white grids (3) lead to a decay rate that is 20 times the rate following exposure to the changing and flashing colored squares of a random grid (2). Random visual stimulation (0) preceded the experiment to start participants at the same baseline.

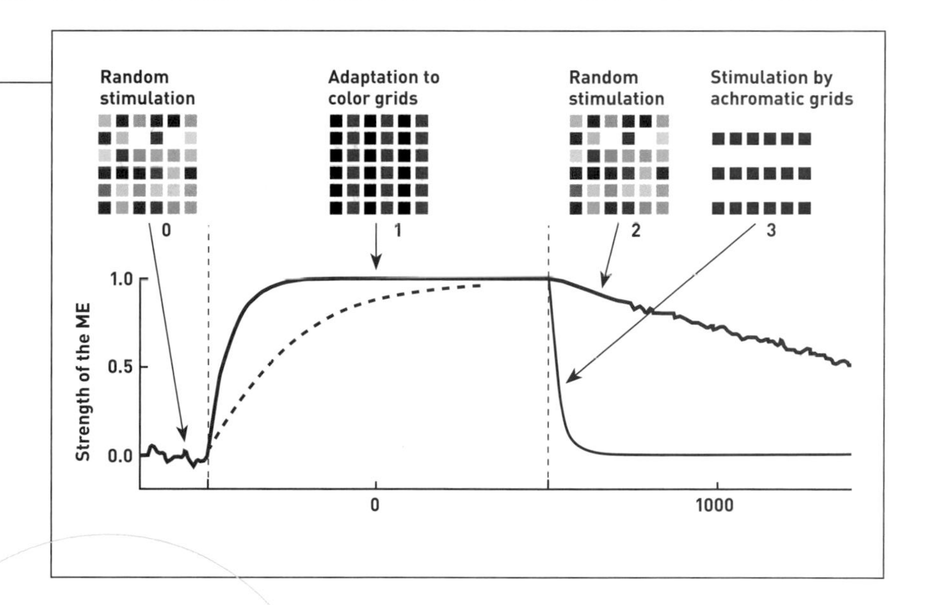

corner. An error-correcting device might interpret prolonged exposure as a deficiency in the visual processing system. The McCollough effect thus would amount to overcompensation among the stimulated neurons.

McCollough's own theory points to color adaptation of edge-sensitive neurons. Her studies indicate that the effect occurs before visual stimuli reach the so-called binocular cells of the cortex's V1 region, where fusion of images from the two eyes begins. ■

The takeaway

Sensations that begin as simple transductions—electrochemical changes in rods and cones—become complex as they get modified in the visual cortex. Neural circuits that detect edges somehow get paired with neurons that process colors, and this creates the long-lasting McCollough effect. The brain possesses mechanisms for recognizing visual elements such as line and color, and it compensates for prolonged exposure to particular stimuli. When it overcompensates, the result can be an illusion.

Focus

your attention on the mounted cowboy in the red shirt in the upper central portion of the watercolor by renowned American painter Bev Doolittle.

Do you see

a realistic, rugged landscape of rocks, trees, and flowing water? Or do you see something more? Give the painting a few more seconds of your time.

What is hiding

among the rocks and trees?

Now turn the page.

What happened

Given the illusion of three dimensions by line, shadow, and perspective, Doolittle's collections of shapes suddenly resolve themselves into human faces.

Leaves, branches, and rocks behind the cowboy coalesce to form American Indians, some of whom gaze upon the cowboy as if their spirits still inhabit the physical world. Doolittle has cleverly incorporated 13 Indian faces in the painting. The painting demonstrates two important phenomena of perception: pareidolia and Gestalt object recognition.

Pareidolia is the human brain's innate ability to recognize patterns, even if they are merely suggested by the weakest of lines and shadows. It causes some people to see faces on Mars or on the carbonized surface of a piece of toast. It doesn't take much for the brain to make this

Bev Doolittle's home page: **http://www.bevdoolittle.net/**
Paintings understood through Gestalt rules of perception.

connection—after all, a "Have a nice day" smiley face is nothing more than two dots and an arc in a yellow circle. Doolittle's work is more sophisticated, but it demonstrates the same principle.

The second phenomenon, which emerged from the work of the Gestalt psychologists, provides a set of rules for understanding how the brain combines the basic elements of visual perception—the edges, lines, and dots—into the whole objects that make up our visual world. Psychologists call these rules the laws of perceptual organization. One law suggested by Doolittle's painting is meaningfulness. Your brain seeks to group things in meaningful, familiar ways. Once it does so, it clings to the new perception. After you see Doolittle's Indians, you cannot un-see them. ■

Pareidolia

The human brain is wired to see patterns in all objects, including random stimuli. This phenomenon, called pareidolia, causes us to find faces everywhere.

Look closer

A train ride in 1911 set psychologist Max Wertheimer on the path to challenging the established structuralist school of perception, which claimed the brain constructs images of recognizable objects out of retinal data.

Wertheimer left his car to get some exercise at a station and bought a toy from a vendor. The toy was a stroboscope, which produced the illusion of movement by rapidly switching views between two pictures. You can see similar illusions today. Consider the lighted headlines that seem to crawl around the electric signboards in Times Square. The letters appear to slide, but they do not. Instead, the illusion of movement occurs when lights flash on and off in sequence.

Wertheimer wondered how the brain could create movement out of two static images. He and his colleagues formed the Gestalt school of psychology—*Gestalt* meaning "form" or "shape." They questioned the brain's ability to decode a mountain of raw data.

Gestalt psychologists believe they can learn much about the mechanisms of visual perception without having to pick them apart, and the whole is greater than the sum of its parts. Confident that mental functions reveal themselves in perceptions as informed by experience, they prefer psychology to physiology. Through experience and experiment, they formed laws of perception such as the following:

It is our task to inquire whether a logic is possible which is not

1. Prägnanz. The German word translates approximately as "good figure," but this phenomenon also is called the law of simplicity. It says the brain's experience causes every pattern of stimuli to resolve itself into the simplest possible structure that would explain the pattern. If you see a brown dog's tail to the right of a tree trunk and a brown dog's head to the left of the trunk, prägnanz causes your brain to perceive them as parts of the same dog.

2. Similarity. This law says that when you see similar objects near to each other, your brain tries to group them. Flocks of birds, all the same color and roughly the same shape, become one group of birds in the mind's eye. Or, if you see rows of squares and rows of circles, your brain groups each shape with its relatives.

3. Proximity, or nearness. The brain groups together objects perceived as near to each other, even if those objects match other objects elsewhere in the visual field. If you see four runners at

piecemeal. Max Wertheimer, address to the Kant Society, 1924

a cross-country meet as they approach the tape, in groups of two separated by ten yards, your mind will likely group the first two and the latter two, rather than one runner from each group.

4. Good continuation. Your mind chooses to see lines as smooth and continuous rather than bending at odd angles. Similarly, your mind chooses to create smooth lines and curves out of arrays of points, rather than random or short, choppy line segments. When you look at the power cords that cross behind your television, computer, and DVD player, you can trace the path of a single cord, even if it crosses over others of the same color and thickness.

5. Common region. When elements appear to be in the same common region of space, your mind groups them together. If you scatter a handful of pennies on the blacktop of your driveway, your brain groups them all together. But draw a chalk circle around every group of three or

All perceiving is also thinking, all reasoning is also intuition, all

four pennies until every penny lies within a circle, and your mind perceives the contained coppers as belonging together.

6. Uniform connectedness. When visual stimuli are perceived as being somehow connected, as by a line, color, texture, or other property, your mind groups them together. Instead of circling the pennies described above, try connecting them, two by two, with chalk lines. You associate the connected pairs.

7. Synchrony. Your mind associates two things happening at the same time. Magicians use this law to their advantage. By sleight of hand, they move two objects very quickly. One object disappears from view, while another object appears in its place. Your mind conflates the two and makes it seem as if the magician has transformed a handkerchief into a bird.

8. Common fate. This law states that our brains group objects that appear to be moving in the same direction. When you see schools of fish swimming in an aquarium, your mind perceives them as one unit—almost like one big fish—if they swim at the same speed and in the same direction. But if you startle the fish, so that some swim to the right and some to the left, your brain turns them into two separate units.

9. Meaningfulness or familiarity. This brings the ideas of the Gestalt psychologists to bear on Doolittle's painting. At first, observation is also invention. Rudolf Arnheim

BREAKOUT

The brain's guides for the eye

Stimulus salience: Environmental characteristics that call attention because of unusual color, contrast, brightness, or orientation, such as a red apple in a pile of oranges.

Previous attention: Familiarity with similar scenes stored in memory such that the eyes scan for expected elements, like the actions of the quarterback after the snap of a football.

Demands of the observer's task: Salience of steps required to complete an action that overrides stimulus saliency, such as attention to the process of making a salad.

Learning from personal experience: The memory-based, cognitive method of examining a scene and increasing salience of stimuli in expected places, such as the greater likelihood of seeing a stop sign on a street corner than in the middle of a block.

you may not have seen the faces in the painting of the cowboy in the woods. However, your brain resolves the visual stimuli of rocks, branches, leaves, and water into familiar images: human faces.

In summarizing the ideas behind Gestalt theory, Wertheimer said in a 1924 lecture, "There are contexts in which what is happening in the whole cannot be deduced from the characteristics of the separate pieces." ■

* The takeaway

Your brain takes shortcuts, formed by experience, to classify and group objects. Gestalt psychologists described these shortcuts as laws of perception. Although technology unavailable to the original Gestalt investigators has begun to probe the mechanisms of perception, Gestalt's holistic laws remain popular because they are fast and easy heuristics of perception. The brain makes quick judgments about whether visual stimuli form familiar objects or group themselves because such judgments usually are correct.

A
B

Study

the checkerboard pattern and the cylinder resting on its right-hand corner.

Note the shadow

cast by the cylinder from some unseen source of light. In the middle of the shadow lies a square, marked B. Higher on the page, outside the shadow, lies another square, marked A.

How

do A and B compare?

Now turn the page.

What happened

A and B are exactly the same shade. You can demonstrate this by isolating the two squares with small pieces of paper. Or, you could lay two bars of gray, the same darkness as square A, across the edges of A and B.

Light bounces off the checkerboard to reach your eyes. Years of experience have trained your brain so that it needs to decide how an object is lighted (or whether it is in bright light or shadow) before it decides the color of the object itself. You know

Visual meaning

Your visual system breaks the checkerboard image down into meaningful components and helps you understand what you're looking at. In so doing, it interprets the gray squares differently, based on context.

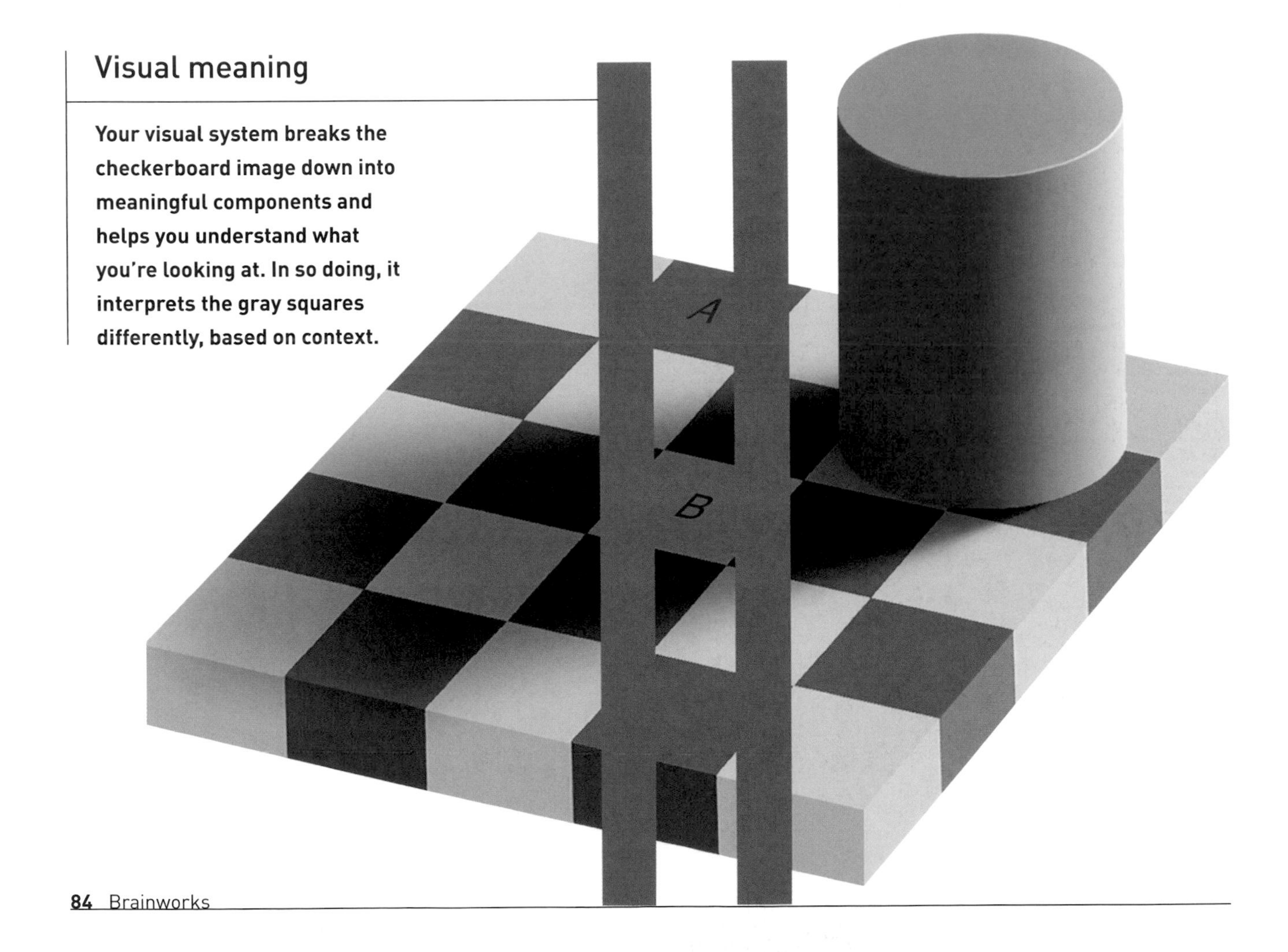

that sometimes a light-colored object in dim light appears darker than a dark-colored object in full sun. But in strong light, objects appear both in shadow and in full illumination. Your brain has learned to compensate for shadows as it wrestles with the issue of an object's "true" color.

In this illusion, called the Adelson checkerboard for its inventor, Edward H. Adelson, your brain makes a mistake. It perceives square B in shadow, surrounded by darker squares. Based on your experience of seeing objects in shadow, it concludes that B's actual brightness, or value, is much lighter. In addition, the cast shadow has soft, fuzzy edges, but the squares' edges on the surface of the checkerboard are sharp. Following Gestalt rules of object recognition, your brain interprets sharp lines as the result of changes in surface color, but it tends to ignore gradual changes in light level, as is suggested by the shadow. The combination of these visual tricks causes the brain to see A and B as different. ■

Likeliest odds

Your brain interprets perception of objects to make the best fit with what it has learned by experience. Lights and shadows can reveal the shapes of fingers and hands—or those of faux animals cast upon a bedroom wall.

Look closer

Your visual system is not a camera; neither is it a photometer. Your brain gathers visual stimuli through the retina and turns them into electronic signals. It then groups and assigns meanings to them based on probabilities, and it uses these probabilities to make sense of the world.

In other words, your brain constructs the likeliest objects it can, given its exposure to light and shadows, lines, movement, and color. Only a few of these meanings are inherited with your DNA. The rest come from your accumulated experience. Without experience to guide you by assigning probabilities to visual stimuli, your brain would have a hard time assigning meaning to things you see.

It's a two-way street. Your brain assigns probabilities to images it sees based on its data bank of visual experiences. It then judges the reliability of new

> Seeing depends on contrast. The retina signals primarily *differences* of brightness. Richard L. Gregory

Penrose triangle

Mathematician Roger Penrose, who popularized his strange triangle in the 1950s, called it "impossibility in its purest form." Although it appears concrete at first glance, it cannot exist in Euclidean space. M. C. Escher incorporated the triangle in some of his mind-bending artwork.

visual stimuli by comparing them to previously assigned probabilities. You recognize a strange new animal as a kind of dog because its qualities seem to match dogs of other breeds you have experienced; you use your memories of dogs to test the probabilities you assign to a new image. If you decide that this new animal is in fact a dog, that information becomes part of your data bank, which judges future visual stimuli of furry, four-legged animals.

Bayesian calculations

In the 18th century, Reverend Thomas Bayes described probabilities as both mathematical ratios and subjective states of mind. Bayes calculated the likelihood of human perceptions in an essay published in 1763 after his death. In the essay, often cited today, he claims that the likelihood of a hypothesis can be determined by examining prior probabilities (previous evidence) and posterior probabilities (the likelihood of new evidence being true if the hypothesis is true). According to Bayes, we can find the odds favoring a hypothesis by multiplying the prior probability by the likelihood of the new evidence, assuming the hypothesis is true, and then taking the ratio of the two numbers.

Neuroscientists believe the brain can store a variety of perceptual hypotheses, each having its own prior probabilities. Very quickly, the brain performs a sort of Bayesian calculation to assign the most likely probability to any new perception.

When you view the Adelson checkerboard, your brain calculates different probabilities for the relative darkness of squares A and B because of prior perceptual probabilities. Two key prior probabilities: objects in shadow appear darker than objects in light, and objects surrounded by light shapes appear darker than objects lacking such contrast.

Now these new facts are added to *your* perceptual experiences. You may call on this information to try to see through future illusions of light and shadow, but chances are you still won't be able to do so. ■

*The takeaway

Lights and shadows help define the contours of objects, and they help your brain visualize objects as occupying three dimensions. Your brain also interprets the colors of objects under varying conditions of light and shadow by applying prior perceptual probabilities. These include the likelihood that objects appear darker if they lie in shadow or against a lighter-colored ground. Illusions such as the Adelson checkerboard exploit these probabilities and create confusion.

Hollow face

Your brain balks at perceiving objects that run counter to experience and logic. An excellent example is the hollow-face illusion, in which a concave face, which does not exist in nature, appears to us as a convex face.

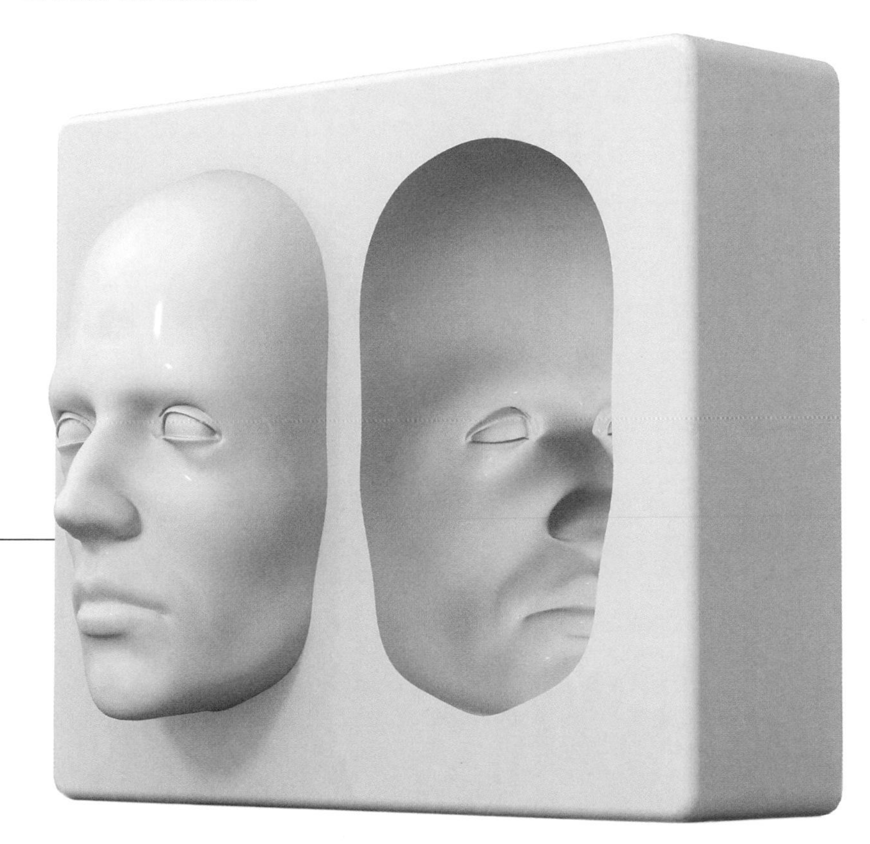

Examine

the painting by Renaissance artist Giuseppe Arcimboldo.

Notice

the detail with which the artist re-creates onions, carrots, and other vegetables. Look how the paint fashions the illusion of a salad bowl through curved lines defining the rim and arcs of white paint suggesting light reflecting off a smooth, round surface.

It's just a painting

of vegetables that Arcimboldo collected. Or is it?

Now turn the page.

What happened

When you rotate the painting 180 degrees, you see a human face. The artist, Giuseppe Arcimboldo, knew intuitively what neuroscientists have long suspected and recently begun pinpointing with physical evidence: The brain actively seeks faces among the chaos of objects it perceives.

Arcimboldo learned his craft under the tutelage of his painter father, and later from Leonardo da Vinci. Arcimboldo executed many conventional portraits of nobles and turned to his tongue-in-cheek, facelike still lifes as a way to express his creativity. ■

Perceiving faces provides advantages for survival. It's important to recognize friends, family, and enemies. So that we benefit from these facial-recognition powers, the human brain essentially operates with an overactive imagination. It sees faces virtually everywhere, which is more important to survival than not seeing the face of someone who could help or hurt us. Pareidolia is much stronger for faces when they appear right side up instead of upside down because that's overwhelmingly how you experience human faces you see in the world. You may have seen a face in the first Arcimboldo image, with the bowl at the bottom of the picture, but chances are your mind first processed the image as a simple collection of vegetables.

The illusion of faces emerging from everyday objects made Arcimboldo famous. Some Renaissance artists specialized in realistic portraits. Others did amazingly realistic still lifes. Arcimboldo combined the two by bringing forth human faces from clusters of ordinary objects. And he didn't limit himself to vegetables. Fruit, fish, flowers, shells, and books also portrayed living people in his still lifes.

Vegetable art

Giuseppe Arcimboldo's 1590 painting "The Vegetable Gardener" strongly suggests a human face, but only when seen right side up. The brain seeks facelike images in this orientation because that is overwhelmingly how faces appear.

Stone face

New Hampshire's Old Man of the Mountain, also known as the Great Stone Face, consisted of five granite formations that suggested a human profile. The formation collapsed in 2003.

Look closer

Neurons that respond to complex visual stimuli tend to be grouped in regions called modules. Modules in the temporal lobe specialize in recognizing particular shapes.

For example, Keiji Tanaka of the RIKEN Brain Science Institute has discovered neurons that respond strongly to lollipop-like shapes that combine a circle and a line, but weakly to either a circle or a line independently. In the 1990s, tests on monkey brains uncovered modules in the inferotemporal cortex that responded strongly whenever the monkeys looked at faces.

Face neurons

Researchers know that the processes of excitation, convergence, and inhibition of photoreceptors make certain neurons in the visual system particularly sensitive to pinpoints of light. They believe something similar underlies the neural machinery that responds to faces. However, it is far more complex to recognize a face as human or, beyond that, to recognize your mother's face, than to acknowledge a circle or a line. Such complexity arises from the brain's trillions of synapses available for processing information.

But which of those synapses are actually firing when we are exposed to images of faces? Hubel and Wiesel's 1950s investigative method of placing sensors on individual neurons would do little to answer this question. They sought simple responses in individual cells, while the complexity of face recognition is spread across entire networks.

Face blindness

From the Greek for "face" and "lack of knowledge" comes *prosopagnosia,* a difficulty or inability to recognize faces. The condition is believed to stem from damage or impairment to the right fusiform gyrus. It sometimes runs in families, suggesting a genetic link.

Fortunately for neuroscientists, a new tool for exploring the brain has become widely available. It allows real-time imaging of modules at work without having to open the skull and insert electrical sensors.

Imaging a working brain

Functional magnetic resonance imaging, or fMRI, works because neurons require more oxygen, delivered by red blood cells, when they are active. Under the right conditions, extremely powerful magnets can detect iron-rich

hemoglobin in the blood. An fMRI detects changes in magnetic patterns in the brain when blood flow to active neural networks increases. In 1997, neural investigator Nancy Kanwisher and her co-workers at the Massachusetts Institute of Technology showed images of faces and other objects to human test subjects undergoing fMRI. They found a small region in the fusiform gyrus, below the inferotemporal cortex, that responded most strongly to faces. They named it the fusiform face area, or FFA. Another region of the temporal lobe, the extrastriate body area (EBA), responds to images of body parts such as hands, legs, and feet, but not faces.

Prosopagnosia

Further evidence of the role of specific brain regions in identifying faces comes from people with a deficiency that interferes with facial recognition. Sometimes the deficiency occurs after a stroke has destroyed a particular neural network. Sometimes the problem exists from birth. Other times, an injury such as the

CASE STUDY

Word blindness

A stroke robbed Canadian mystery writer Howard Engel of his ability to read and recognize common objects. Strangely, it did not affect his ability to write. Neuroscientists call this rare condition *alexia sine agraphia.*

Engel's world seemed normal until he grabbed a newspaper one morning in 2001. Headlines looked as if they had been written in Cyrillic or Korean, he recalled. At first, he thought someone had played a practical joke. Then he realized something had gone wrong in his brain.

Hospital tests confirmed damage to Engel's visual cortex, including neural networks that normally activate in recognition of the alphabet, colors, faces, and objects. Engel said apples and oranges looked foreign to him. Yet, at a nurse's suggestion, he found he could write sentences.

Engel's misfortune demonstrates the brain's compartmentalization of functions. Reading activates the inferotemporal cortex, crucial to the recognition of printed words by their letters and shapes. It works with any language, from Japanese to Sanskrit. As writing emerged relatively recently in human history, neuroscientists think this brain region evolved to recognize topographic shapes. The brain's plasticity transferred those recognition skills to letters on a page.

Engel regained some reading ability by tapping that same plasticity to recruit neurons from other brain regions. He began to trace letter shapes with his tongue on the inside of his mouth, a way to incorporate tactile and motor sensations to the task of word recognition.

> My attention goes to the rest of you—your voice, shirts, your pants. I am not very good at . . . faces. Oliver Sacks

bump of a traffic accident or a shell fragment in battle weakens a portion of the brain or renders it useless.

When a deficiency interferes with your ability to recognize faces, the resulting condition is called prosopagnosia. The term comes from Greek words meaning "face" or "mask" and "lack of knowledge," so sometimes the condition is referred to as face blindness. It can range from mild (having difficulty recognizing a familiar face) to severe (not realizing that an arrangement of features constitutes a face at all). There is no cure, but people who have it can train themselves to recognize people through cues other than their faces, such as clothes, height, and hair color. The German neurologist who coined the condition and wrote its defining paper in the 1940s told of a wounded soldier's method of identifying a particular nurse. The soldier had suffered a brain injury and could not recognize any faces, including his wife's. However, he could pick out one of his nurses because she had blazingly white teeth that shone when she smiled.

Voices, which provide auditory stimuli processed independently of visual sensations, are a valuable aid to people with prosopagnosia.

Living with face blindness

In his 2010 book *The Mind's Eye*, neurologist Oliver Sacks reveals that he has dealt with some degree of face blindness his entire life. He sometimes has difficulty recognizing his students, his colleagues, and even his own image in a mirror.

He's not alone. Researchers at Harvard believe about 2 percent of people have some degree of face blindness. Given the complexity of the visual processing system, perhaps that's not so surprising. ■

Portrait parts

American painter Chuck Close, in a self-portrait, has face blindness, yet he paints recognizable images of people. "I commit them to memory," he says, "by flattening them out and carefully scanning them." A face is just "colored dirt on a flat surface."

* The takeaway

Modules in the temporal lobes specialize in object recognition. A small module called the fusiform face area responds to faces and facelike patterns, even in a bowl of vegetables or dots and a curved line. It also can pick out one particular face from the thousands of people you have known in your lifetime. You might even recognize a loved one rendered skillfully in rutabagas, potatoes, onions, and carrots.

Chapter 2

KING

Perception of the world creates a rich tapestry

> Memory is the diary that we all carry about with us. Oscar Wilde

of sensations in the brain. But if humans experienced

the world as animals do, these sensations would evaporate like dew on a summer day. People would live in a never-ending *now,* with no idea of where they have been or where they might go. It is the memory of representations of the world that gives humans the unique ability to think about the past, present, and future. • Memory turns the brain into a virtual time traveler. Reflecting on past events serves as the foundation for decision-making as the brain considers present and future. Imagining the possible outcomes of contemplated actions helps the brain set goals and visualize pathways to achieve them. These mental functions have a strong reason for existence: They served humanity's ancestors by helping them survive, breed, and pass their emergent memory skills to their children. Early humans who remembered that a particular snake

Memory paths

Light stimulates the retina (1), which transmits impulses to the visual cortex (2) and on to the frontal cortex (3), which coordinates their immediate use by other parts of the cortex (4, 5). The hippocampus (6) and other areas of the medial temporal lobes (7) turn short-term memories into long-term ones and activate stored memories into working memory (8).

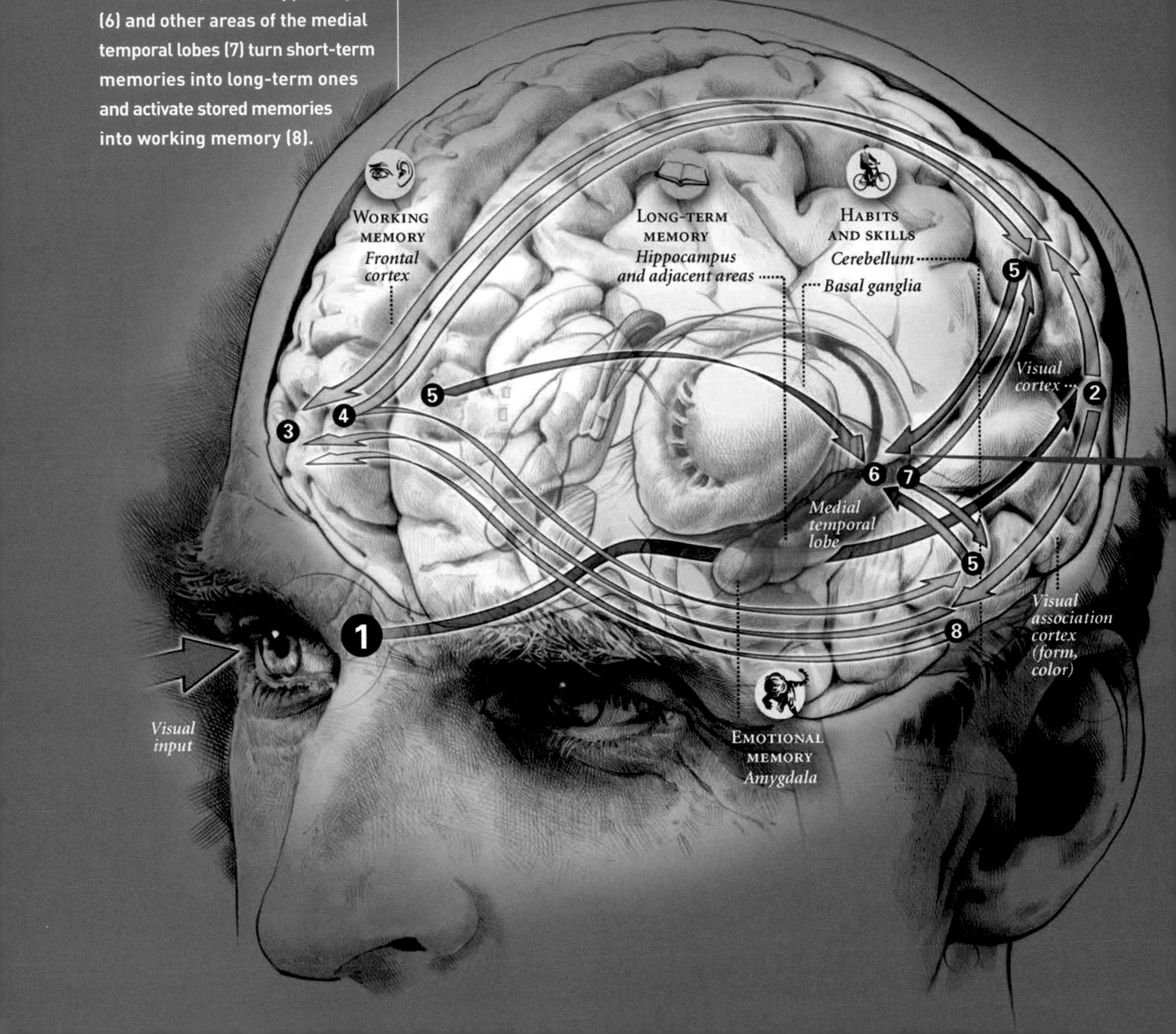

killed the people it bit would avoid that snake in the future, thus raising their own chances for survival. And those who remembered how to find food in a drought had an advantage over those who could not.

Modifying behavior in response to memories of past events is learning. Modifying behavior through memories that prompt imagination of the future is planning. Both learning and planning emerged as the forebrain expanded and began processing sensory data in a more sophisticated way.

Fact

Over your lifetime, your brain may store as many as one quadrillion (a million billion) bits of information in long-term memory.

Memory

Scientists who study memory find it a tough nut to crack. Neural circuits responsible for memories of the appearance of a rose, the weather on your wedding day, and the importance of the frontier in American history lie scattered throughout the brain. Experts have found some brain regions to be particularly crucial, however. The hippocampus, so-called because it resembles a seahorse (*hippos* is Greek for "horse"), plays a significant role in emotion and memory. Lying deep in the forebrain, it receives sensory data from the senses and integrates them into a single experience.

Much of the information reaching the hippocampus gets ignored and never enters memory. At any moment, your brain is flooded with millions of bits of information; you pay attention to some. Often it is by choice. When you watch television, your

Juggling act

Sensory data fill your brain but do not overwhelm it, thanks to conscious and unconscious processes that allow you to focus attention where it's most needed.

If we remembered every thought and sense . . . our brains would

Memory aids

You can remember the past (retrospective memory) or hold information for future use (prospective memory). A string tied around the finger acts as a prospective memory cue.

brain processes sights and sounds onscreen and ignores other stimuli. Sometimes, however, sensations demand your attention, as when the roar of a wild animal or the screech of tires jerks your focus from other things. Again, both kinds of attention have advantages for survival. Sustained concentration allows you to perform complex tasks, from cooking dinner to communicating important ideas to others. And involuntary, physical reactions to unexpected stimuli may save you from harm.

According to Jeff Hawkins, a neuroscientist who invented the Palm Pilot, your brain solves problems by retrieving answers from memory. You compare the world you know with your memories of it, and then you make predictions. Where is your car? What's on your plate for breakfast? When your predictions fail, you revise your understanding of the world. Prediction, carried out by the neocortex, is "the foundation of intelligence," Hawkins said. Prediction matches what your senses tell you is happening with what you expect to happen, based upon your memories. Your predictions then form the foundation of your decisions.

Some memories last less than a minute. Some last a lifetime.

Working memory

Working memory, a type of short-term memory, holds sensations for a few seconds. That's a good thing. If every sensation you experienced over a lifetime remained forever clear in memory, you'd have little room for other thoughts. The brain holds in working memory only the sensations it needs to interface smoothly with the world. This phenomenon allows your brain to operate more efficiently. You remember the last few words you spoke or the last few ingredients you put in the stew so you may complete the tasks you have begun. You remember stimuli so you may react to them—dialing a phone number after you hear the dial tone, paying the right amount of cash when told your bill, or giving the right answer when the teacher calls your name. Then working memory clears, and you move on.

One way to think of working memory is as a desktop, either on your computer or on a table. There you store bits of digital information or paper that you need in order to perform a particular task. Too much clutter—too many programs running at once, or too many papers in too many stacks—interferes with your ability to get the job done. At some point, the computer crashes or papers tumble to the floor. For example, the brain can hold only a few random numbers in working memory.

Long-term memory

Long-term memory is more like a filing cabinet. Some documents from a computer screen or tabletop get filed for later use. Information is more likely to make the transition from working to long-term memory if you pay particular attention to it, repeatedly try to remember it, or associate it with strong emotions. These memories are useful not as much for navigating through the moment as for making good choices and succeeding at life.

Scientists have wrestled with demystifying the mechanisms that place information in memory. One school of thought, propounded by neuropsychologist Karl S. Lashley, argues that memories form through so-called mass action of the brain. Another school, founded by neuropsychologist Donald O. Hebb, proposes the localization of some memories via alterations in cell assemblies through changes in neuron-to-neuron connections. According to this theory, one cell assembly might hold the name of an object, another its appearance, and another the sound it makes.

Scientists have found evidence of memory-related changes in neuronal connections in a variety of sources. In 2007, University of California, Irvine researchers found expanded synaptic connections in rats' hippocampi after the rats learned new tasks. And scientists at University College London discovered that taxi drivers on average had larger

Memory and age

Childhood amnesia robs most adults' memories of events before age three or four, likely because a child's brain isn't fully formed. In adulthood, memory declines gradually.

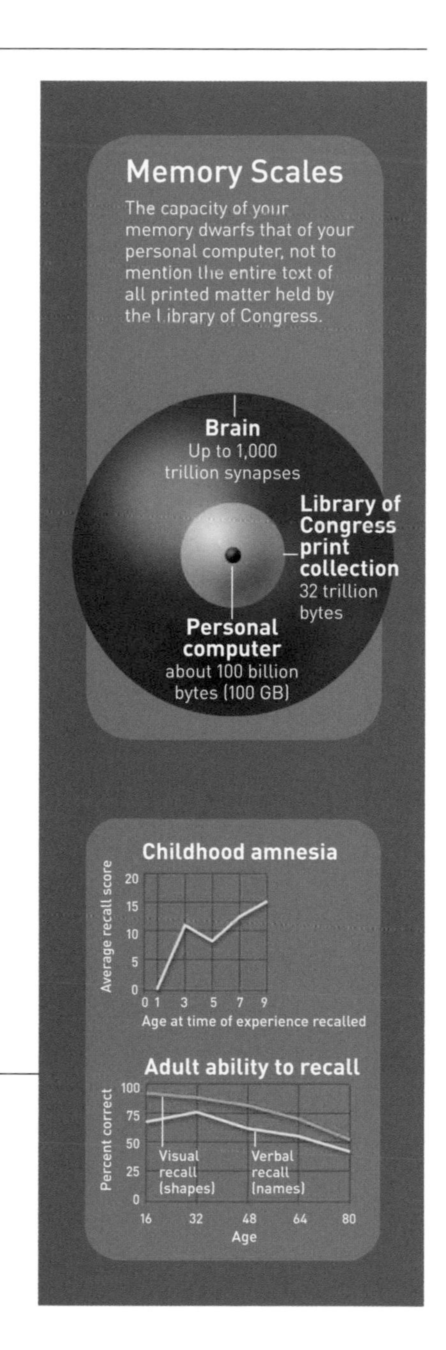

hippocampi than other people and that this brain region grew as the cabbies spent more time on the job. "The hippocampus has changed its structure to accommodate their huge amount of navigating experience," said Eleanor Maguire, who led the research team.

Damage to the hippocampus impacts its ability to form new memories. This was vividly demonstrated by a patient known as H.M., who had brain surgery to relieve his seizures. Removal of portions of his hippocampus and temporal lobes caused H.M., who died in 2008, to live for decades in the present moment. He was unable to move information from his mental desktop to his long-term cabinet.

Taxi training

By studying maps of London, a man trains to become a taxi driver. Tests on London cabbies have revealed enlarged hippocampi, indicating enhanced memory about addresses and street navigation.

Memory stages

Neuroscientists refer to the stages of memory as encoding (initial integration of information), storage ("filing" of the memory), and retrieval (fetching of the memory from storage for current use).

Encoding starts with the gathering of stimuli when we are at attention. Factors including elaboration and emotion enhance this process.

Elaboration is the association of one bundle of information with other bundles. Associating a word with an image encodes two kinds of information—verbal and visual—and strengthens the overall encoding. Thus, a common memory technique is to use mental images of words as mnemonic devices. Our senses of smell and taste also are powerful memory encoders, as they are you lose the functions. Joseph LeDoux, *Synaptic Self*

the only senses directly wired into the hippocampus. French novelist Marcel Proust discovered this when he took a sip of tea and nibbled a shell-shaped confection in 1911. "No sooner had the warm liquid mixed with the crumbs touched my palate than a shudder ran through me and I stopped, intent upon the extraordinary thing that was happening to me," he wrote. Transporting Proust's current mind back to childhood, the taste had evoked memories of a garden and Sunday morning.

Emotion can enhance encoding when we associate sensory stimuli with the feelings they generate. If the emotions are strong, such as the terror of watching the Twin Towers collapse on September 11, 2001, so-called flashbulb memories, which include surprising amounts of detail, may form. A small, almond-shaped brain region known as the amygdala, which is crucial to our fight-or-flight response to sudden noises, is believed to mediate the encoding of memories with emotional content.

Storage . . . and learning

Storage maintains encoded information. Some memories are known as explicit or declarative, meaning the conscious mind can call them up. These include memories of facts and language, called semantic, and memories of personal experiences, called episodic. Other memories are unconscious, or nondeclarative.

Very young children appear to develop semantic memory before episodic memory. At two years, children can demonstrate memory of events from seven months earlier. Until they reach about age four, however, they neither include themselves in memories nor appreciate the measurement of time, a skill that appears to relate to episodic memory. Episodic memory's relatively late development also explains why most people have few, if any, memories of their lives as babies. Meanwhile, as the child's brain learns to understand time, it also begins to remember spatial relationships. Children's first drawings reveal space as they know it,

Fear response

Strong emotional reactions, such as a burst of intense fear that activates the amygdala, strongly encode memories. If you were ever in fear for your life, you likely remember the moment in detail.

not as they see it: Every object is seen in full view without overlap, with no evidence of scale, and with shapes of objects rendered independently of viewpoint. Only after learning visual cues of three-dimensional space does a child begin to draw an image that more accurately represents light falling on the retina.

The beauty of neurons is that because they are not physically connected, their terminals easily shift to forge new relationships. When neurons make new connections, or when they strengthen or weaken existing connections, the brain adds information—it learns. Scientists now know that when you learn words in a foreign language or master the D-minor chords on a guitar through repetition, that information gets stored through changes in your neural circuitry. This process isn't entirely clear, as memory formation involves at least 117 kinds of molecules, but researchers believe it involves changes in transmission of neurotransmitters and their receptor sites. In 1999, scientists found that by genetically engineering mice to enhance their N-methyl-D-aspartate (NMDA) receptors, the mice's learning and long-term memory improved.

So, the brain reshapes itself as it makes memories. But those new configurations don't stay constant. Your memories change as you retrieve them.

Flashbulb memory

A woman holds photographs she took of the events of September 11, 2001, in New York City. So-called flashbulb memories occur when especially vivid, emotional events occur. They defy erosion and continue to make you "stop and think," as the *New Yorker* cartoon, opposite, suggests.

Retrieval

The act of recalling a memory activates the same neural pathways activated by the initial experience. Sensory memories spring from the hippocampus.

Emotional content gets added from the brain's limbic system. Other neural circuits may add aural content, linguistic content, or spatial arrangements. When connections become strong, the memory grows stable enough to exist without the intervention of the hippocampus—which explains why H. M. could remember events from before his surgery.

Retrieval is not a simple process of pulling information from a file and then refiling it. Because the brain constantly looks for patterns, retrieval of long-term memories calls up not only stored data, but also other information associated with the memory. Some of the new associations stick to the original memories as they return to storage. In other words, the act of retrieving a memory changes it. It is not difficult to envision the origin of false memories, particularly under the power of suggestion. Things you heard or saw today could get mixed with events from another day.

Fact

Elderly people have more difficulty learning word lists—and forget them more quickly—than college students do.

Memory manipulation

In 2000, New York University researchers Joseph LeDoux and Karim Nader discovered that they could erase memories by tinkering with retrieval and storage. They taught rats to associate a loud noise with an electric shock. Then they injected the protein inhibitor anisomycin at the precise moment that the rats' brains retrieved the connection between the stimulus and the response. Conventional wisdom suggested that when the anisomycin left the rats' bloodstreams, their long-term memories linking shock with noise should have remained intact. Instead, when the inhibitor blocked the rats' ability to remember the fear-based memory in that one specific circumstance, the original memory evaporated. Similar tests demonstrated that rats' episodic memories of spatial designs also could be wiped clean.

It is not too wild to picture a future scenario that mirrors the plot of the 2004 science-fiction movie *The Eternal Sunshine of the Spotless Mind:* Human memories get altered at the touch of a button. ■

"It sort of makes you stop and think, doesn't it."

Flexible memory

Experiment 2.1

Would you

make a good witness in court? Examine these photographs of the automobile accident. Suppose the crash happened while you were walking nearby, and someone later asked you to testify about what you saw.

How much

could you recall about the scene? How accurate would your memories be?

Take a few moments

to study the photographs.

Now turn the page.

? What happened

Without turning back the page, answer these questions: Did you see broken glass on the ground? Headlights smashed? Skid marks?

The answer is, it depends.

Imagine your friend Sally calls to tell you that her friend, Tim, has been in a traffic accident. You realize it was the one you just saw. Sally says Tim feels so stupid: As he tuned his car radio and texted on his mobile phone, he sped through a red light and *crash!*—he smashed head-on into another car.

That's when you tell Sally you saw the accident. "Could you give a statement about it?" she asks. You agree.

Now return to the opening questions. What about those shards of broken glass, smashed headlights, and skid marks?

Return to the photograph.

If you recalled seeing broken glass or some other detail that wasn't there, don't be too hard on yourself. It's a common reaction in the studies performed by psychologist Elizabeth F. Loftus. She examines how memories become distorted by post-event information, including the process of recalling those memories over time. Loftus has demonstrated that the presence of detail, expression of confidence, and emotional content of memories does not necessarily indicate that events happened as we remember them.

When pressed by an interrogator, particularly one using leading language, witnesses to crimes and accidents can form false memories. Loftus found that when witnesses were asked how fast two cars were traveling when they "smashed" into one another, they recalled higher speeds than when asked about two cars that "hit" one another. Furthermore, the verb *smashed* increased the percentage of people who were certain they saw broken glass. ■

False memory

The introduction of the word *smashed* versus the word *hit* may influence your visual reconstruction of an accident.

HIT

SMASHED

Mix and match

Elizabeth Lee McShane Stuart, peering through an opening on a glass tabletop, was best known for her elaborate, custom-made wooden jigsaw puzzles, which lacked the usual placement clues of ordinary puzzles.

Look closer

Many people think of memory as an audio/video recorder: The brain tapes an event and stores it in a library to be retrieved, played, and refiled.

According to this view, every playing yields the same "story." Memory is much more complicated than that. Psychologist Daniel L. Schacter provides a new metaphor: Memories are like jigsaw puzzles. When you recall an event, you bring together pieces from many storage bins in the brain. Sometimes the pieces get mixed up.

Articulating a memory creates what Loftus calls a story-truth: The brain creates a satisfying story about actual events. Such story-truths help people understand their personal history and create identity.

Just as a novelist plays with plot while drafting a chapter, or a puzzle assembler sometimes tries out the wrong pieces, the brain may alter, fill in, and delete portions of a story-truth during the act of remembrance.

Circuits of memory

Alterations occur because of how memories are formed, stored, and retrieved. Scientists believe memories are formed and stored when neurons forge links to create new circuits. Although they do not fully understand these circuits, researchers have done much to trace the links for declarative memories, which require conscious retrieval. Declarative memories include episodic ones, which contain information about

CASE STUDY

Blind imagination

In your mind's eye, picture someone close to you—mother, father, friend. Close your eyes, and conjure the image.

Easy, right? For a patient known in scientific literature only as M.X., it's an impossibility. His imagination sees only blackness.

The troubles began for M.X. shortly after treatment for blocked coronary arteries in 2005. M.X. felt a "reverberation" in his head. He apparently suffered a brain injury that erased his ability to form mental pictures.

Neuroscientists Adam Zeman and Sergio Della Sala examined M.X.'s brain. They tested him against a control group of ten similar men, all architects in late middle age. Scans showed no differences in how the brains of M.X. and the architects activated facial-recognition networks when they were shown faces. However, Zeman and Della Sala noted a profound change when they named famous people and asked the men to imagine the matching faces. In the control group, every circuit for recognition of faces—except for those involved in direct visual perception—became active. M.X.'s visual circuits remained quiet.

And yet Zeman and Della Sala found that M.X. could describe details of Edinburgh landmarks and facial features of famous people. He also could look at an object composed of ten cubes and determine if it was the same as a similar object seen from a different angle. Most people given that task mentally rotate the objects in their mind's eyes, but M.X. had no such ability. That led the researchers to suspect that when M.X.'s brain lost its ability to create mental images, it may have rerouted visual information to other circuits below the level of consciousness. Perhaps his brain found a method not involving visual pathways to solve visual problems. Zeman and Della Sala dubbed M.X.'s condition blind imagination.

personal history (Where did I go to school?), and semantic ones, which contain general knowledge (Who was President in 1944?). Circuits for particular kinds of semantic memories, such as names of fruits and vegetables, appear to occupy specific regions throughout the cerebral cortex. Episodic memories, on the other hand, appear to rely on localized connections to the medial temporal lobes and the hippocampus. Scientists believe independent neural networks process the what, when, and where of episodic memories, as well as sensory details stored in regions associated with their original sensations—for example, visual data stored in the visual cortex. The act of remembering assembles the pieces. For example, you draw together the color, shape, and size of your dog, along with the feel of her wet fur and her enthusiasm, when you recall a game of fetch on a California beach.

Psychologists refer to the connection of stored information

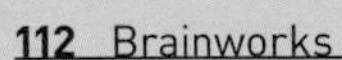

> Are we aware of our mind's distortions of our past experiences? In most cases, the answer is no. Elizabeth Loftus

fragments as memory binding. The binding is strong when you have a flashbulb memory, which links your cognitive experiences to a powerful emotional experience encoded by the amygdala. Adding contextual cues—you recall how your mother cut her finger as she made Thanksgiving dinner during a terrible snowstorm—increases memories' strength by linking them with other memories.

Searching, searching . . .

What goes on in your brain when you try to recall images? Harvard psychologist Stephen Kosslyn is fascinated by how people answer the question "What shape are a German shepherd's ears?" He has found that the brain stores patterns of visual data that are activated when seeking the answer. Hence, you create a mental image of a German shepherd, right or wrong, to "look" at its ears. The same thing happens when you try to recall how many windows were in the house where you grew up, or the shape of the crunched automobile at the beginning of this experiment. Brain scans of people seeking to answer questions about shape, color, and other details reveal activation of many of the same neural circuits used in perception of the original

Visual memory

What shape are a German shepherd's ears? To answer the question, you likely conjured an image of the dog for your mental inspection.

Stephen Kosslyn's laboratory: **http://isites.harvard.edu/icb/icb.do?keyword=kosslynlab**
Articles and audio files and descriptions of books for further reading.

The bits of information that we recover from the past are often

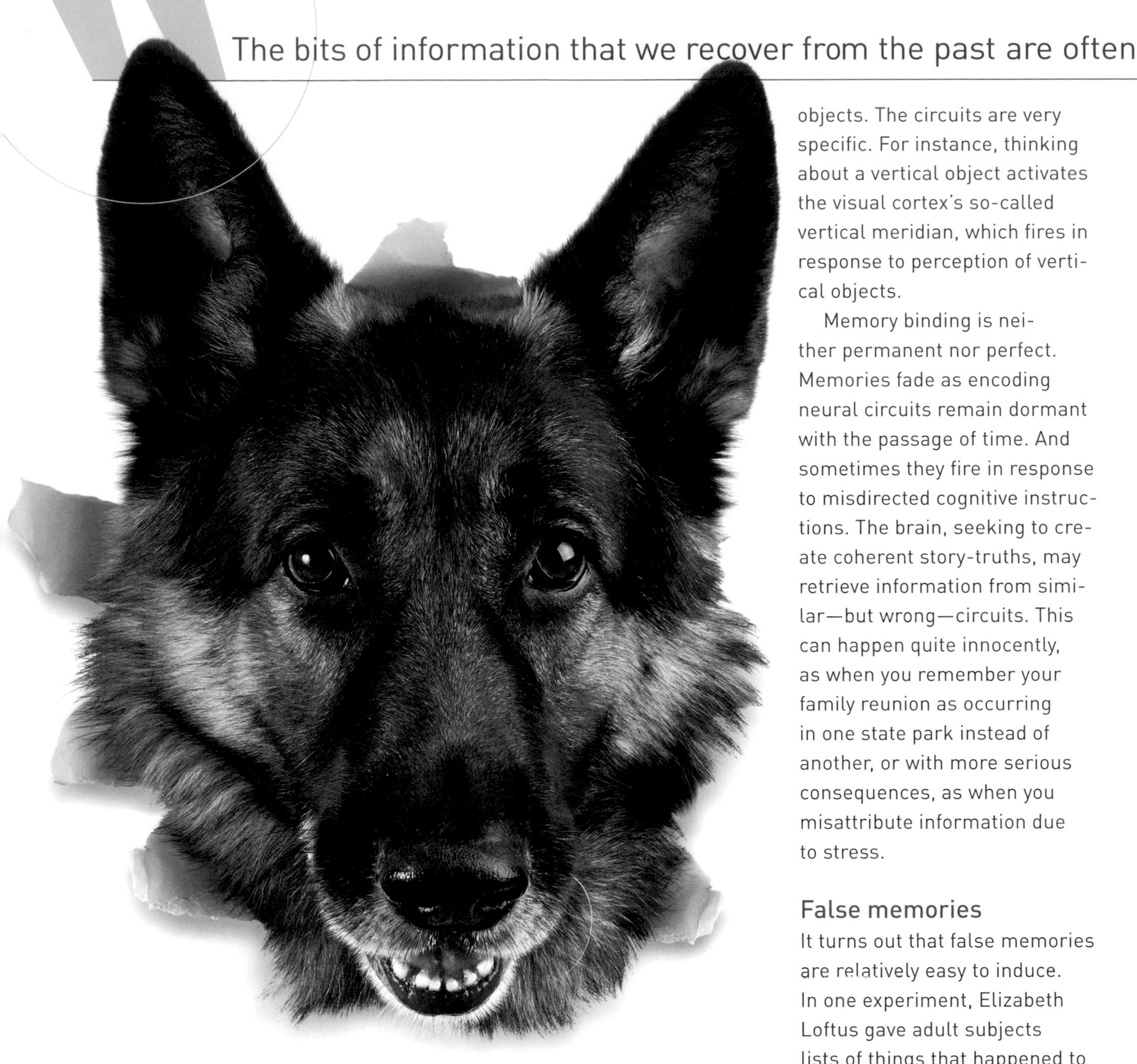

objects. The circuits are very specific. For instance, thinking about a vertical object activates the visual cortex's so-called vertical meridian, which fires in response to perception of vertical objects.

Memory binding is neither permanent nor perfect. Memories fade as encoding neural circuits remain dormant with the passage of time. And sometimes they fire in response to misdirected cognitive instructions. The brain, seeking to create coherent story-truths, may retrieve information from similar—but wrong—circuits. This can happen quite innocently, as when you remember your family reunion as occurring in one state park instead of another, or with more serious consequences, as when you misattribute information due to stress.

False memories

It turns out that false memories are relatively easy to induce. In one experiment, Elizabeth Loftus gave adult subjects lists of things that happened to

Elizabeth Loftus's research: **http://faculty.washington.edu/eloftus/**
Influential articles by the groundbreaking memory expert.

them as young children. Family members had provided the information to ensure that the lists contained real events that the subject likely would remember. However, Loftus added a lie to the roster: The subject had gotten lost at a mall. About a quarter of the participants said they remembered the false event, and many gave elaborate details. In another experiment, Loftus showed the subjects advertisements for a Disney resort that featured the cartoon character Bugs Bunny. One in six subjects later recalled having met Bugs at Disneyland. That was clearly false, as Bugs Bunny is a Warner Brothers cartoon.

"People's memories are not only the sum of all that they have done, but there is more to them: The memories are also the sum of what they have thought, what they have been told, what they believe," Loftus said. "Who we are may be shaped by our memories, but our memories are shaped by who we are and what we have been led to believe." ■

BREAKOUT

Streams of memory

Short-term or working. Lasting a few moments, it holds something in mind long enough to complete a task or to make sense of a sentence by bringing together its beginning and end. The brain may convert it to long-term memory.

Long-term, procedural. Implicit memory of how to perform an action unconsciously. Sometimes called muscle or motor memory. Once acquired, it allows such actions as typing and guitar playing without thought of where to move the fingers.

Long-term, episodic. An explicit memory in which conscious thought recalls personal experiences.

Long-term, semantic. An explicit memory in which conscious thought recalls learned facts, such as those acquired in school lessons.

* The takeaway

Memory is fluid, not fixed. The power of suggestion can form false memories, and even memories expressed with confidence can contain errors. The brain creates memories that make sense of the world and a person's place in it. Such fluidity calls identity into question. If identity is the sum of what you remember, who are you if memories change? It appears that for good or ill, you become the person you imagine.

1

Thread
Pin
Sharp
Point
Sewing
Eye
Thimble
Prick
Thorn
Hurt
Haystack
Injection
Syringe
Cloth
Knitting

2

Bed
Tired
Rest
Awake
Wake
Dream
Snooze
Blanket
Slumber
Doze
Snore
Nap
Drowsy
Peace
Yawn

Remembering Words

Experiment

2.2

Carefully study

the words in each list. Start with column 1 and then move to column 2.

Now turn the page.

What happened

Which of the following words appeared in the lists? Choose yes or no for each word. Now flip to the previous page.

The above experiment is known as a DRM test. That's short for the names of the three researchers most closely associated with the procedure: James Deese, who developed the test in the 1950s, and psychologists Henry L. Roediger III and Kathleen McDermott, who rediscovered it and gave it to college students in the 1990s. Roediger and McDermott found what they call "remarkable levels of false recall and false recognition" among undergraduates asked to identify words they were told to remember from lists of 15 related terms. So, for example, when presented with words such as *thread, sewing,* and *knitting,* as many as half the students remembered, incorrectly, that the list contained the word *needle*. Similarly, they swore that the second list of words, which included *blanket, doze,* and *pillow,* contained the word *sleep*. Neither word appears on the previous page. Harvard psychologist Daniel Schacter reports false memory rates in such experiments as high as 80 to 90 percent among hundreds of students who attend his lectures.

According to McDermott, urging people to be on their guard does little to change the results. "Warning them about the possibility of illusory memories does not permit people to control their thought processes and avoid having them," she said. "It's clear that people have difficulty suppressing false memories. The key questions now are how and when are these mistaken memories generated and can they be avoided?" ■

Look closer

False memories of unseen and unheard words arise from associations. Your mind associates all the words from the first list with *needle* and all the words from the second list with *sleep.*

"We believe that the concept of 'sleep' is aroused through activation spreading through a semantic network while people listen to the list words," Roediger said. Activation occurs both externally, as test subjects hear the words, and internally, as they associate the words into a coherent structure.

In the mid-1990s, Schacter studied whether positron emission tomography (PET) scans could determine whether the brain reacts differently when accurately and inaccurately attesting to a word's appearance on the original list. If such differences could indeed be demonstrated, a PET scan would function as a truth machine: It would go lie detector tests one better by unmasking people who are unaware that they are lying.

Schacter's results were no such breakthrough. He found that test subjects' brain activity followed similar patterns whether they gave true or false answers about the word lists. In both cases, strong electrical activity in the frontal lobes appeared, with weaker signals in the inner regions of the temporal lobe and the hippocampus. At first, whispers of differences tantalized Schacter's research team. A portion of the frontal lobe related to the monitoring of memories seemed more active during false reports. True memories also sparked a bit more activity in a region of the left hemisphere associated with word sounds. These results disappeared when the tests were repeated under more rigorous conditions, however.

The power of association

Canadian psychologist Allan Paivio has found that a hierarchy of associations heightens memory. The brain remembers objects more strongly than it remembers pictures of the same objects. It also remembers pictures of objects better than it remembers

BREAKOUT

Common memory errors

Transcience. The erosion of memory over time.

Absentmindedness. The weak formation of memory cues through lack of attention.

Blocking. Thwarted efforts to retrieve known information.

Misattribution. The assigning of memory to a false source.

Suggestibility. The creation of false memories through suggestion.

Bias. The editing of memories in light of present conditions.

Persistence. Repeated recall of events preferred to be forgotten.

Source: Daniel L. Schacter, *The Seven Sins of Memory*

CASE STUDY

Jean Piaget's mistake

Swiss psychologist Jean Piaget (1896–1980) is remembered as a titan in the field of child psychology. His work on the stages of mental growth, which occur as children learn to adapt to their environments, is considered seminal. He demonstrated how young minds think in vastly different ways than mature ones.

Yet even such a pioneer in the workings of the young brain fell victim to its quirks. For a long time, Piaget held a false memory that he fervently believed to be true.

"I can still see, most clearly, the following scene, which I believed until I was about fifteen," he said. "I was sitting in my pram, which my nurse was pushing in the Champs Élysées, when a man tried to kidnap me. I was held in by the strap fastened around me while my nurse bravely tried to stand between me and the thief. She received various scratches, and I can still see vaguely those on her face. Then a crowd gathered, a policeman with a short cloak and a white baton came up, and the man took to his heels. I can still see the whole scene, and even place it near the tube station."

Years later, the nurse joined the Salvation Army. As a result of her religious conversion, she decided to come clean with Piaget's parents. She told them she had faked the entire episode, including the scratches on her face. She returned the gold watch that the parents had given her for her bravery.

Piaget apparently had heard the story as a child and then had stored the false account in his mind as a "true" memory. Even as an adult, when he knew the memory was untrue, he had trouble sweeping it to the dustbin of mental error. He called it "a memory of a memory, but false."

words for objects, which after all have no concrete connection to the objects themselves. (Why is a rose called a rose?)

The power of memory increases when you link words to images. That's because the encoding of such memories occurs across a variety of neural networks, including those for language and vision. This explains the efficacy of a memory technique that dates to ancient times. Called the Roman room for its description by Roman poet

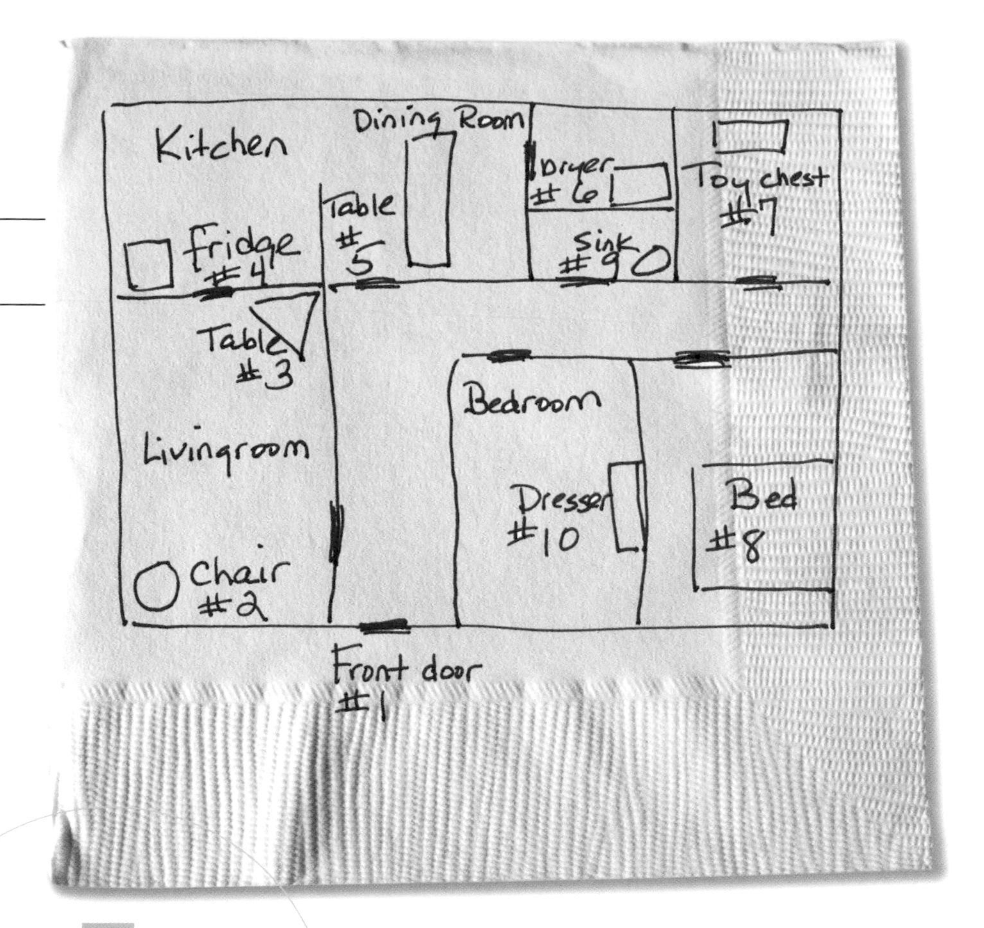

Roman room

Via an ancient memory technique, you imagine memory cues along a familiar path. You could place atomic elements or presidents in your rooms.

Cicero, it works by linking words and phrases to images. If you create a concrete image for a word and place it in a particular location, you are more likely to remember the word. For example, you can remember the French word for father, *père,* by picturing a yellow pear wearing a smoking jacket and holding a baby. If you string words and images together in a narrative way, such as going from room to room, you can remember long chunks of text.

Schacter demonstrated that visual associations boost the accuracy of DRM tests. When he showed pictures while subjects heard words from a DRM list—a picture of butter accompanying the pronunciation of the word—the association proved decisive. Subjects correctly said they had heard a particular word only when they could recall seeing its image. ■

The takeaway

Hearing or seeing lists of words activates neural networks of perception and cognition. Your brain weaves the two together to make coherent associations. As a result, you may develop false memories. Inside your brain, false memories quickly become just as true, in your estimation, as real ones. If that's the case for something as simple as 15 words on paper, consider how likely it is for more complicated events, such as crimes or car accidents.

4926

57843

02685

953456

684712

1537923

7430673

89314289

85371067

639818531

9278369708

The power of seven

Experiment **2.3**

Start

at the top of the list of numbers.

Look

at the number, try to memorize it, and then cover it with your hand.

Immediately say the number out loud, one digit at a time, without pausing.

You would say the first number as "four-nine-two-six."

Continue

along the list of numbers. Note the point at which your memory starts to falter in your attempts to recall the entire number.

Now turn the page.

What happened

Did the number that beat you begin with 8? The strings of digits that start with an 8 are also eight digits long. Chances are, those numbers, and the longer ones below, gave you trouble.

For a long time observers have noted that the mind can manipulate only so much information at one time. You may have noticed that when a friend tells you a new phone number and you immediately try to write it, you likely forget it if your friend asks you a question while you scribble. Your brain can't handle the overload. But at what point does the brain rebel against handling too much at once?

Pioneering cognitive psychologist George Miller conducted a series of experiments and proposed the answer in 1956 in the *Psychological Review.* "My problem is that I have been persecuted by an integer," reads his paper's famous opening. The mysterious numeral is evident from the article's title: "The Magical Number Seven, Plus or Minus Two."

Miller found most people could keep about seven new pieces of information in memory. Some could hold a little less, some a little more. Through subsequent experiments, researchers have tweaked Miller's numbers in certain situations and added insights into how different kinds of information affect memory's boundary. Some have questioned Miller's finding that active memory can hold about seven items, regardless of whether the data exist as numerals, letters, words, or some other form. Nevertheless, the end of Miller's article asks an intriguing question: Is it just coincidence that there are seven seas, seven ancient wonders, seven deadly sins, and so on? ■

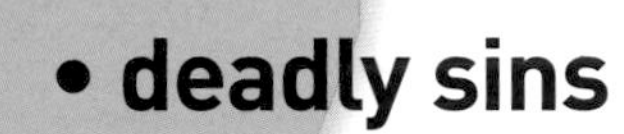

- **deadly sins**
- **ancient wonders**
- **brides for 7 brothers**
- **days in the week**
- **dwarfs**
- **phone digits**
- **hills of Rome**

Look closer

The brain has more than one kind of memory. Long-term memories include general facts and the meanings of words, as well as personal events such as your wedding day. The former are called semantic memories, and the latter are episodic.

You need semantic memory to communicate ideas. You need episodic memory to have identity, as who you are depends upon what you remember.

Procedural memory, also known as muscle memory, is long-term as well. It develops when complex movements become routine. When you first try to ride a bike, your frontal lobes consciously send signals to muscles to keep the wheels turning, the frame pointed in the right direction, and the bike upright and balanced beneath you. After a while, control of these movements shifts from forebrain to cerebellum, and muscle memory lets you ride without conscious thought.

You also have the kind of ephemeral memory that lets you hold about seven things in mind. Scientists have discarded the term *short-term memory* in favor of *working memory* in order to indicate its active, crucial role in performing mental work. It resides in the prefrontal lobes.

In the 1970s, British psychologist Alan Baddeley proposed that working memory acts as a temporary platform for reasoning, mental arithmetic, problem solving, and so on. By holding bits of information while adding to them or comparing them with others, working memory lets the brain forge new insights. Working memory allows you to carry on conversations by holding someone's words in mind while you formulate a response.

Baddeley developed a three-part model of working memory. It consists of what he called a central executive, a phonological loop, and a visuo-spatial sketch pad. The central executive acts as cognitive boss, directing attention at the phonological loop and the visuo-spatial sketch pad, as well as controlling the interface with long-term memory. When you read something that you must remember for an exam, your central executive works with the other parts of working memory to help you understand it and then attempts to direct it to long-term storage for later retrieval.

The phonological loop acts like a piece of recording tape.

Memory storage

Long-term memories aren't stored like socks. Retrieval jumbles them—just as your socks may lose mates, pair with others, or disappear.

Exploratorium home page: **http://www.exploratorium.edu/explore/mind/**
San Francisco's museum of science, art, and human perception.

CASE STUDY

The man who forgot

When Henry Molaison died in a Connecticut nursing home in 2008, his *New York Times* obituary called him the most important patient in the history of brain science. It was a description he would have found utterly forgettable—just like everything else since 1953.

His troubles began at age nine when a bicyclist slammed into him and he sustained a serious head injury. That led to seizures and, by age 27, incapacitating blackouts.

Believing that Molaison's troubles arose from localized injuries, a surgeon removed much of his hippocampus and the inner surface of both medial temporal lobes. The procedure did indeed halt the seizures. It also destroyed Molaison's ability to form memories.

Molaison became famous in scientific literature as patient H.M., his name reduced to initials to maintain privacy until death. He could recall World War II, his name, and other things he learned before his surgery. Afterward, he could hold numbers in active memory for a few seconds and carry on a conversation as long as his working memory could keep a topic in mind. But nothing stuck.

Psychologist Brenda Milner, who studied Molaison for many years, said, "Every time I walked in the room, it was like we'd never met."

Milner's studies revealed three key facts: First, the brain treats memory as distinct from cognition, motion, perception, and other neural functions. Second, the brain stores working and long-term memories separately. Third, regions crucial to converting working memories into long-term ones lie in the medial temporal lobes and hippocampus, demonstrating that some brain functions are highly localized.

Molaison's condition made clear the close connection between memory and identity. Without new memories, Molaison remained essentially the same person.

Muscle memory

Don't think—do. When you call upon procedural memories to make a well-practiced ski jump, shoot a free throw, or type on a keyboard, your forebrain forms the proper motions of the new physical skill. With practice, motor skills become automatic, their memories move to the cerebellum, and you act without thinking.

As you read the numbers at the beginning of this experiment, their sounds ran through your mind like snippets of tape. According to Baddeley, sounds in the loop decay after about two seconds, an idea that mirrors Miller's conclusions about limits to active memory. The visuospatial sketch pad holds images for manipulation, but like bits of data in the phonological loop, they decay rapidly.

All three parts may work together. When you multiply two 2-digit numbers in your head, the phonological loop holds the numbers themselves. The step-by-step process of multiplication

occurs in the central executive. Images of the numbers in columns activate the visuo-spatial sketch pad. Once you have the answer, the process evaporates from memory unless it is so important that you move it to long-term storage.

And thank goodness. Imagine the clutter if everything you ever held in working memory remained filed away for all time. ■

* The takeaway

Working memory places limits on the amount of information your brain can manipulate or, if you so desire, direct toward long-term storage. With training, however, you can associate and sequence information to store long, complex strings. For a moment, you can remember a seven-digit phone number. But by organizing information into chunks, you can eventually commit to long-term memory the Gettysburg Address or the first 50 digits of pi.

AE592
E5
FOR ALL DEBTS,
Treasurer of the United States.
SERIES
100
100
DOLLARS

How we choose

Experiment **2.4**

For the following scenarios, choose (**a**) or (**b**).

In case 1, you're given $1,000 and must choose

(**a**) a 50-50 chance that you'll get $1,000 more, or nothing more, or

(**b**) a certain gift of $500.

In case 2, you're a doctor choosing between two treatments for an Asian disease entering the United States. The disease is expected to kill 600 people.

If you choose (**a**), 400 will die.

If you choose (**b**), there is a one-third chance of zero deaths and a two-thirds chance of 600 deaths.

Now turn the page.

What happened

Did you pick (b) for both cases? Most people do.

Daniel Kahneman, a Princeton University professor, and his research partner, Amos Tversky, reported that 84 percent of respondents picked (b) for the first case. When physicians heard the second scenario, 78 percent selected (b).

Now suppose the two cases were presented this way:

In case 1, you're given $2,000 and asked to choose (a) a 50-50 chance you'll lose $1,000, or nothing, or (b) a sure loss of $500. In case 2, option (a) would save 200 of the 600 infected people, while option (b) would present a one-third chance of saving 600 and a two-thirds chance of saving nobody.

Given these descriptions, 69 percent of respondents chose (a) for case 1, and 72 percent of doctors chose the safer strategy, (a), for case 2.

If you do the math, you'll find a surprise. The two versions of the monetary and "Asian disease" scenarios offer equally probable outcomes. In the two versions of case 1, the odds are the same that you will end up with $1,000 or $2,000, if you choose that route, versus the sure bet of $1,500. In the two versions of case 2, the odds are the same that you'll get zero or 600 deaths, if you choose to gamble, or 400 deaths (or 200 survivors) if you take the sure thing.

The only difference is framing. When the problem is framed as a potential loss—either money or human lives, in these cases—the brain exhibits a bias toward taking risks. ■

Which hand?

Evolutionary pressures for survival bias the choices you make—which hand holds the prize and which the penalty? When you've got something good, you don't want to risk losing it. The pain of a potential loss dwarfs the pleasure of an equal-size gain.

Look closer

These experiments helped make Kahneman the first psychologist to win the Nobel Prize in economics.

His key finding was that the brain weighs the potential pain of a specific loss *more* than the potential pleasure of an equal-size gain. Some psychologists estimate that losses pack twice the mental impact of gains, which raises interesting questions about the brain's biased wiring.

Gains versus losses

Merchandisers manipulate this phenomenon every day with sales pitches. They emphasize gains over losses, even if the mix of the two remains constant. Consider yogurt. If it's advertised as 95 percent fat free, that means it's 5 percent fat. But advertising something as 5 percent "bad" doesn't appeal to a brain that favors a product that's 95 percent "good."

This flies in the face of thousands of years of philosophy. Plato, writing in the fourth century B.C., pictured the mind as a chariot being pulled by a pair of strong horses. The charioteer represents reason. The noble, well-behaved right-hand horse represents the spirited mind, which honors positive attributes such as self-preservation. The ugly, wild, left-hand horse represents the "ignoble breed" of negative attributes, such as greed, lust, and gluttony.

According to Plato, the rational mind fights to keep the wild horse of negative emotions from bolting. Centuries later, Sigmund Freud echoed Plato by dividing the brain into the horse of the id and the rider of the ego: "The horse provides the locomotive energy, and the rider has the prerogative of determining the goal and of guiding the movements of his powerful mount towards it." In both metaphors, good decisions happen when reason controls emotion.

Pleasure versus pain

In the case of economic choices, Kahneman and Tversky's experiments punctured the armor surrounding what philosopher John Stuart Mill called *Homo economicus* ("economic man"). This species supposedly acts to maximize well-being amid constraints and choices. According to Mill, people assign measurements to pleasure and pain. Think of pleasure and pain, or gains and losses, as weighing various numbers of bricks.

Angels, demons

Faced with making a quick decision, you often rely on emotional memory to guide your actions. Did you feel good or bad about an earlier choice?

CASE STUDY

Monkey economics

Economics has long been considered unique to humanity. "Nobody ever saw a dog make a fair and deliberate exchange of one bone for another with another dog," said Adam Smith, the 18th-century philosopher whose writings explained laissez-faire capitalism.

Yale University economist M. Keith Chen's work has called Smith into question. In doing so Chen also has found evolutionary evidence for the endowment effect, in which people overestimate the value of things in their possession. The effect runs counter to economic theory by injecting irrational behavior into Smith's model of decision-making based on rational, narrow self-interest.

Chen introduced tubes of peanut butter and frozen juice bars to a laboratory group of capuchin monkeys. Both foods can only be eaten slowly, so the monkeys were forced to consider the value of the food and the possibility of exchanging it instead of gobbling it down.

Overall, when given a choice, the monkeys split 60 percent to 40 percent in preferring peanut butter to juice bars. Yet possession of one or the other introduced strange economic behavior: Some monkeys who preferred peanut butter were given a juice bar and refused to trade it; some who preferred a juice bar were given peanut butter and refused to trade it. It was as if the monkeys felt a risk in giving up what they had.

Other researchers theorized that the endowment effect stems from an evolutionary bias in favor of conservation of vital resources. They discovered that monkeys showed no such effect for toys made out of bones and ropes. Conclusion: Toys are toys, at least for monkeys. But food is vital, and trading it poses risks. Further along in evolutionary development, humans have evidenced the endowment effect with pens, coffee mugs, and virtually anything else that can be possessed.

These bricks are then placed on a scale. You opt for choices that most heavily tilt the scale toward happiness. If the scale must tip the other way, you pick the option that leads to the smallest of all possible pains or losses. If Mill were correct about human rationality, you would look at the choices in Kahneman and Tversky's scenarios and make the same choices regardless of how they are phrased, because the scale always balances the same. Instead, people seem to swing irrationally from one choice to the other merely on the basis of how the choice is framed.

> We're not aware of changing our minds even when we do change our minds. Daniel Kahneman

Status quo bias

Your brain balks at giving up or changing things you see as yours. This influences everything from your wine collection to your stock portfolio.

Emotion versus reason

Scientists now know that "irrational," emotional choices often prove superior to those made on the basis of strictly rational assessment. Emotional prediction acts as a kind of shortcut to good decisions. Faced with a choice, you may make a quick decision based on the memory of how you felt after making similar choices. "Let's say you are trying to make a complicated decision," says neurologist Antonio Damasio. "If you try to do a cost-benefit analysis, it may take you forever to decide whether to do A or B. However, if you have previously been in similar situations, and if you have been either rewarded or punished by the choices you made in those situations, then emotional memory may help you with your current choice."

In truth, emotion and reason work best when they work together. In his book *How We Decide,* Jonah Lehrer writes, "Sometimes feelings can lead us astray and cause us to make all sorts of predictable mistakes. The human brain has a big cortex for a reason."

Endowment effect

One apparently irrational economic quality is ownership. It slants the perception of value.

> Be wary of intuitions, especially intuitions about how your own mind works. Daniel Simons

"Let me pick up an ashtray from a dime-store counter, pay for it and put it in my pocket—and it becomes a special kind of ashtray, unlike any on earth, because it's mine," says Gail Wynand in Ayn Rand's novel *The Fountainhead.* Kahneman and Tversky established the existence of an "endowment effect," in which people like Wynand place greater value on something after acquiring it. If asked to put a dollar value on an object before possessing it, they do so, but when given the object, they decline to part with it unless they are given substantially more money than they previously named. Similar, seemingly irrational behavior takes place as soon as test subjects see themselves as "owning" the money in the opening experiment.

Status quo bias occurs when people disfavor change. People hold on to bad stocks for too long, or they refuse to trade them for investments that they would have purchased if starting portfolios from scratch. They refuse to part with personal items, such as a family heirloom, even when offered a very high price. Or, on a television game show, they keep the prize they have rather than trading it for something unknown.

The economic brain

In 2008, Brian Knutson of Stanford University used fMRI to examine test subjects' brains as they performed tasks designed to provoke the endowment effect. They bought products, received products and sold them, and then made choices between cash and other products. The scans found increased activity in

BREAKOUT

Decision points

As you make decisions, researchers have found that where you end up often depends on where you start.

Try this: You walk into a store to buy a radio for $25. Someone tells you the same radio is available a few blocks away for $20. Would you walk to the other store to save the money?

Now this: You walk into a store to buy a TV for $500. Someone tells you the same TV is available a few blocks away for $495. Now would you walk? According to the experiments of behavioral economist Richard Thaler, the answer is yes to the first scenario and no to the second, despite your saving $5 each time. The reason for the difference is that most people weigh the $5 against their starting point, the amount they're spending. It's a large percentage of the radio but a fraction of the TV.

three brain regions. The first, the nucleus accumbens, which has been linked to predictions of monetary gains, became more active when the subjects dealt with preferred products they bought and sold. The second, the mesial prefrontal cortex, linked to revising an initial assessment of a likely gain, declined in activity as prices rose during buying, but increased activity as prices rose during selling. The third, the insula, linked to prediction of monetary loss, varied in response to dealing in preferred products; researchers found its activity levels could predict how strongly a person felt the endowment effect. These reactions appear to have sprung from an evolutionary survival strategy of minimizing the risk of bad losses.

Gum ball count

Make your best estimate of the number of gum balls. (You'll start with an anchor and move up or down.) Turn the page for the answer.

> Status quo, you know, is Latin for "the mess we're in." Ronald Reagan

Anchoring and adjustment

Another example involves what psychologists call anchoring and adjustment. According to this concept, you decide to make estimates of unknown things by starting with a preliminary estimate and then shifting it based on additional information.

Kahneman and Tversky showed how an initial estimate can bias decisions. They asked students to estimate the percentage of African countries that are members of the United Nations. Each person began by spinning a wheel that produced a number between 1 and 100.

The researchers then asked the students if their estimates were higher or lower than that number. Finally, the students stated what they estimated to be the true percentage. When the spin of the wheel turned up a ten or less, the students' average estimate was 25 percent. When the wheel turned up 65 or more, the students' average estimate was 45 percent. Even though the wheel's number had no logical connection to the final estimate, it still influenced what the students decided.

The concept of anchoring finds its way into purchasing behavior. Used car dealers sometimes begin negotiating a sales price by starting at an artificially inflated number and then dropping to the price they want. Buyers think they've saved money and gotten a bargain. ■

The answer

244

The takeaway

When you make a decision, there's more than your power of reason at work in your brain. You're influenced by emotion, and that quite often turns out to be a good thing. You're also subject to a peculiar system of neural wiring that makes you averse to risks that might lead to losses. In fact, apparently the result of evolutionary survival strategy, your behavior makes losses feel much, much worse than gains of similar dimension.

The error of easy memories

Experiment **2.5**

Choose

which element of the following pairs is more likely to happen
to someone close to you:

Murder versus suicide
Death by shark bite versus death by dog bite
Home burglary versus identity theft
Death by allergic reaction to peanuts versus death by accidental poisoning
Abduction of a child by a stranger versus death of a child in a pool
Death by terrorism versus death by aircraft accident

Now turn the page.

What happened

The less likely event of each pair is on the left-hand side.

Fear factors

Heart disease:	1 in 6
Cancer:	1 in 7
Stroke:	1 in 28
Motor vehicle accident injuries:	1 in 85
Accidental poisoning:	1 in 139
Occupant in car accident:	1 in 272
Assault by firearm:	1 in 300
Injuries as a pedestrian in accident:	1 in 623
Rider in motorcycle accident:	1 in 802
Accidental drowning:	1 in 1,073
Smoke, fire, or flames:	1 in 1,235
Firearm discharge:	1 in 5,981
Excessive heat:	1 in 6,174
Murder:	1 in 18,000*
Hornets, wasps, or bees:	1 in 62,950
Lightning:	1 in 81,701
Being bitten or struck by a dog:	1 in 119,998
Any Earth movements:	1 in 153,597
Airplane accident:	1 in 354,319*
Choking on food:	1 in 370,035*
Fireworks accident:	1 in 1,000,000*
Shark attack:	1 in 300,000,000*

Numbers vary slightly from year to year. These odds, from the National Safety Council, represent American deaths in 2006, except for those marked with an asterisk, which represent deaths in 2001.

Suicides in the United States are more than twice as common as murders. Deaths from ingesting peanuts represent a fraction of the thousands from unintentional poisoning. And in an average year (not 2001), deaths from aircraft accidents dwarf those from terrorism, which normally total a few dozen.

Decision-making gets swayed by memory. Specifically, the ease with which you can recall particular facts influences how common you think those events are. It's simple mental arithmetic. The more often things happen, you reason, the more you remember them. And the more often you store such facts and events in memory, the more likely you are to call them to mind. Ergo, easily recalled memories point to the truth.

But memory is fickle. Memory search patterns find some things more easily, a concept known as salience. And some things get encoded strongly into memory because of their vividness. Boring facts don't usually create strong memories.

In the case of the paired examples on page 139, some events receive more prominent news coverage than others, and thus we are more likely to see and recall them. Shark attacks get bigger headlines than dog attacks. And while airplane fatalities do get reported, relatively few occur among the commercial airlines, and, for good or ill, they fade more quickly from the news than deaths inflicted by terrorists. Studies of the frequencies of news stories about particular causes of death have shown an almost perfect correlation with survey subjects' estimates of the frequencies of those fatal events. ■

Look closer

People use only a handful of heuristics, or mental shortcuts, when attempting to judge the frequencies of events.

When you rely on the ease of recall to judge an event's frequency, you employ the so-called availability heuristic. You particularly use it when you don't have concrete evidence to help fashion an informed judgment. For example, if you wanted to estimate your risk of dying in a plane crash on any given flight, you would need to know a host of details. How old is the plane? How many miles has it been flown? How often have qualified mechanics checked it? What are the weather conditions at the site of takeoff and landing, as well as at every point in between? Who is flying the plane? What is the pilot's experience and mental state? And so on.

You lack knowledge of the many variables that would form the basis of a true calculation of risk. So, you use what's at hand: memory. You are more likely to ponder recent airplane crashes you remember and build upon that information to make your best guess. Chances are, if you recently read about a plane crash with multiple fatalities in a newspaper or saw a video report on the television or Internet, you would have strong, readily recalled memories of the crash story on the news. You then would overestimate your chances of dying. If you are the rare person who had a loved one die in a plane crash, your personal, emotional connection would make the memories even more influential. On the other hand, virtually everyone knows someone who has died in a vehicular accident. Yet the fear

The sound of salience

Words that start with a *t* are more salient because you are accustomed to paying attention to a word's first letter, rather than its third.

taser satisfied taste

gotten bottle terrible

of flying outweighs the fear of the much more dangerous act of driving. Car accidents don't register very high on the national news agenda.

Vividness and salience

If you rely on memory to make imperfect judgments, it's no surprise that factors enhancing a memory's encoding skew your judgment even further. Vividness is one such factor. Take the case of your deciding which college class to take as an elective. You could pull the course evaluations filed anonymously by all of the students in the three or four classes you are considering, tabulate them, and choose a class that consistently posts a high average score. Or you could ask a friend who took one of the classes last year. You likely would give more weight to a single friend's angry description of a class he hated than to a mountain of data based on hundreds of happy students. The friend's story is more vivid, more memorable—and more likely to sway your judgment. Advertisers take advantage of this phenomenon. A television commercial containing footage of a luxury car paired with cute babies or woodland animals strikes a memorable, emotional chord. The viewer supposes the manufacturer emphasizes safety or "clean" emissions, whether that's true or not. In a similar vein, the news can distort. Televised sports programs that broadcast baseball highlights in the summer show a lot more homers than routine fly balls. Result: viewers' overestimation of the frequency of home runs. (Current average: two per major league game.)

You also probably fall victim to salience. That is, you become biased toward the effectiveness of your method of searching for memories. If you were asked which are more common, English words starting with the

take mottled tennis tied hot less potato

CASE STUDY

Feeling no fear

The availability heuristic can increase everyday fears by prompting you to overestimate the frequency of risky activities. If you fear to walk alone at night, you don't—or you do so with trepidation.

But what would it be like to be literally fearless?

Scientists have studied just such a woman for more than 20 years. They published an account of her life in the journal *Current Biology* in 2010. To preserve her anonymity, they called her S.M.

S.M.'s background emerged through her stories and diaries. Once, as she was walking through a city park at night, a man pulled a knife and held it to her throat. S.M. didn't panic. She noticed the sound of a nearby church choir and calmly told the man he would have to go through "God's angels" to kill her. The attacker then became the one who panicked, and he let her go. S.M. returned to the park the next day.

Researchers led by Justin Feinstein at the University of Iowa recently put S.M. to the test, exposing her to scary movies, terror-inducing haunted houses, and snakes and spiders that she handled at a pet store. She took it all in fun. She said that while she feels other emotions, she simply has no fear. The research team found no evidence to contradict her.

Feinstein blames brain damage. S.M., now in her forties, was born with a rare condition that gradually destroyed her amygdala, a brain region associated with feeling fear.

The study of S.M. could open doors to treatment of people overwhelmed by negative emotions, such as veterans afflicted with post-traumatic stress syndrome, or PTSS. In a separate neurology study, researchers found that soldiers who suffered damage to their amygdalae were less likely to develop the disorder.

letter *t* or English words having a *t* as their third letter, you likely would choose the former. Words that start with a *t* are more salient because you are accustomed to paying attention to a word's first letter, rather than its third. Because you can remember so many of them, so easily, you likely reason they must outnumber the harder-to-remember words with *t* as their third letter. You also would be wrong.

Correcting for bias

Vividness and salience vary by person. What you most readily call to mind might not match another person's heuristic.

Making choices can be problematic when you base them primarily on the availability heuristic, but the risks decrease when you gather information from multiple sources. That gives everyone's heuristics the opportunity to cancel

Risk assessment

The availability heuristic causes you to overestimate risks through your ability to call to mind certain low-risk but highly publicized dangers. Big news impacts both memory and choice.

each other. One friend may tell a horror story about buying a particular model of car, while a second and third friend may give equally passionate accounts of love affairs with their reliable wheels. The more you talk to people, the more you know.

Scientists have found strength in numbers. Refuting those who sneer at the quality of decisions reached by committee, financial analyst Paul Johnson discovered that groups naturally balance individuals' availability heuristics. In 1998, when he asked a large group of students to predict Academy Award winners in 12 categories, the group's picks pointed correctly to eleven winners. No individual correctly picked more than nine winners. ■

* The takeaway

Judgments become flawed if they place too much stress on information easily called to mind. You probably overestimate the likelihood of events that stick in memory, particularly if they are vivid or your search method accesses them easily. If you live according to the so-called availability heuristic, you may lose focus on life's real dangers. You grow afraid to fly in a plane . . . and therefore hit the much more dangerous highway.

The paradox of choice

Experiment **2.6**

Imagine

going into a gourmet food store and seeing this display of varieties of jam.

Pick one

you think you might like to buy.

Now turn the page.

What happened

How difficult was it to choose? You're more likely to buy a jar of jam if 6 varieties are available instead of 24.

So said researchers Sheena S. Iyengar of Columbia University and Mark R. Lepper of Stanford University. They had two people dress as store employees and prepare two tables for display, one with 6 jams and one with 24. They switched the tables every hour so that as the day progressed, customers would randomly encounter one or the other. Customers could taste as many jams as they wished. They received a coupon, but they had to go to the jam aisle to pick out a jar for purchase. That way, no matter which table they encountered, they would be exposed to all varieties of jam before selecting one for purchase.

The large display of jams attracted more customers than the smaller display. However, only 3 percent of the people exposed to the large array purchased a jar. A whopping 30 percent who went to the smaller table picked out a jam jar and bought it.

"These findings are striking," wrote Iyengar and Lepper. "Certainly, they appear to challenge a fundamental assumption underlying classic psychological theories of human motivation and economic theories of rational choice—that having more, rather than fewer, choices is necessarily more desirable and intrinsically motivating."

Having too much choice appears to impair decision-making. This may be because extra choices force the decision-maker to put forth greater mental effort. That effort may not be seen as worth the gain, or the decision-making process may detract from enjoyment after purchase. ■

Look closer

Decisions define your life. You make them all the time, although you probably don't think about them.

When to get up, what to have for breakfast, what to wear, whether to read the paper or catch the news and weather on the morning TV, what time to leave for work . . . all contribute to the characteristics that define you as unique. Existentialist philosopher Albert Camus summarized this idea by posing the rhetorical question "Should I kill myself or have a cup of coffee?" All of life is choice.

Your brain does not make choices in a vacuum. Instead, it weighs alternatives in relation to each other, to memories of previous experiences, and to expectations of future actions.

Modern life has brought blessings and curses to the process of choice. On one hand, technology has made many choices seem easier and more convenient. If you want to make supper, you could choose to heat a frozen dinner or order takeout, and then leave the leftovers in the fridge, instead of choosing the hard labor of cooking and baking from scratch every night. On the other hand, mass production has made possible so many more choices in every aspect of life—which school to attend, which job to pursue, which phone service to use, which car to buy, which of hundreds of styles of shoes to wear—that the sum total of choices can create stress. Author Barry Schwartz, a professor of social theory and social action at Swarthmore College, found that while shopping in an electronics store, he could mix and match stereo equipment (including speakers, tuners, CD players, and so on) to create 6.5 million different stereo systems. He felt overwhelmed by such numbers.

Got the blues

Slim, easy, or relaxed fit? Button or zipper? Boot cut or tapered?

> There's only one problem with this assumption of human rationality: it's wrong. Jonah Lehrer

A simple pair of jeans

Schwartz felt compelled to write a book, *The Paradox of Choice*, when he set out to buy a pair of jeans. He hadn't bought any for a long time, so he was stunned by the choices in the clothing store. Buying jeans became a nearly daylong chore.

"Before these options were available, a buyer like myself had to settle for an imperfect fit," he wrote, "but at least purchasing jeans was a five-minute affair. Now it was a complex decision in which I was forced to invest time, energy, and no small amount of self-doubt, anxiety, and dread."

As Schwartz points out, absolute freedom of choice does not necessarily create happiness, as classic, free-enterprise economists once believed. It is true that choice is *intended* to lead to enjoyment. The act of comparison can make each option seem worse. We assume we want choice, but when we have it, we are not always happier.

BREAKOUT

Stressless decision-making

Choose when to choose. Some decisions require serious consideration of multiple variables, and some do not.

Increase your satisficing and decrease your maximizing. Perfect choices, like other forms of perfection, seldom if ever exist in the real world.

Think less about opportunity costs. These are the benefits of the choices you reject. Unless you are concretely upset by a clearly bad purchase, stick with what you have bought. And if you make decisions you cannot reverse, force yourself to ignore the options you turned down.

Acknowledge that pleasures change over time. The thrill of a new car fades into the satisfaction of a reliable automobile.

Don't compare yourself to others.

Accept, and even embrace, constraints on your decision-making. Reduce the number of future decisions.

Source: Barry Schwartz, *The Paradox of Choice: Why More Is Less*

Maximizers and satisficers

Some people choose to accept only the best. Schwartz calls these people maximizers. They invest much of their time and energy comparing options to make sure they choose the things they consider the finest. The search is difficult and not very rewarding. After all, to be sure of making the ultimate choice, a maximizer must take on the impossible task of considering all options. And after making a choice, a maximizer is likely to reconsider and have "buyer's regret."

A more psychologically satisfying strategy for making choices is to choose to be happy with excellence rather than perfection. In the 1950s, Nobel Prize–winning psychologist Herbert Simon coined the term *satisficing* to refer to the process of factoring stress, time, money, and other variables into decisions. Satisficers make very good choices—albeit not the absolute best ones—because their brains find the optimal balance between the pleasure of finding the right choice and the pain of the extended selection process.

CASE STUDY

Life without emotion

When doctors removed a small piece of the frontal lobe of a patient named Elliot, they took away more than a tumor. They cut out his ability to decide.

After the surgery, Elliot remained bright. But when making a choice—Where to eat lunch? Which pen to use?—Elliot froze. He deliberated pros and cons without end. After losing his marriage and his job, he visited neurologist Antonio Damasio in 1982.

Damasio noticed no sadness or other emotions in Elliot. Physiological tests confirmed Elliot had lost his emotional life.

As Damasio investigated similar patients, he discovered that emotion arises in several brain regions. However, the orbitofrontal cortex, right behind the eyes, proved crucial. Damage there erased emotion and took effective decision-making with it.

Damasio demonstrated his findings with a card game. He and colleagues at the University of Iowa recruited six players with orbitofrontal damage, including Elliot, and ten without damage. The players got $2,000 in play money, were given some simple rules, and were told to choose among four decks to flip a card. Cards provided money or penalties. Two "good decks" had been stacked to provide small immediate returns but larger payouts over time. Two "bad decks" did the opposite.

The control group quickly developed hunches about decks being good and bad, and they favored the good decks. Three of the six brain-damaged players eventually concluded two decks were bad but kept choosing them anyway. None of the six showed physiological signs of turmoil when flipping cards from bad decks, but the control group did.

Damasio argues that emotional memory underlies choice. Remembering the payout and pain of previous decisions may be the root of "gut feelings," he said.

Destiny is not a matter of chance; it is a matter of choice. It is

Voltaire: **SATISFICER**

Henry VIII: **MAXIMIZER**

The bargain-hunting brain

In 2006, to measure the pleasure and pain of choices, a team of psychology researchers from Stanford, Carnegie Mellon, and the Massachusetts Institute of Technology recorded the brain responses of 26 young adults hooked up to fMRI scanners. Each of the volunteers received $20 and then was shown a series of pictures of the kinds of items people often purchase as gifts. These included fancy chocolates, DVDs, clothes, and books. After an image had been projected for a couple of seconds, the researchers flashed its price, reduced to fit the participants' total budget of $20. The volunteers had four seconds to click a *yes* button or a *no* button on each of 80 items, shown one at a time. At the end of the experiment, the volunteers had an average of 23 items and the researchers had their brain data.

The results mirrored the brain scans for the endowment effect in Experiment 5. If the volunteers liked an item and its price, the fMRI detected increased action in their nucleus accumbens. But items perceived as overpriced excited their insula while dampening activity in the mesial prefrontal cortex. According to the researchers, the brain weighs how much it likes the item against its impact on the buyer's limited resources. Credit cards, they say, dull the pain by delaying its consideration until some future date.

Real-world choices

The brain scans of the pleasure and pain of purchasing decisions took place under extremely controlled conditions. Participants had no opportunity to window shop, to clip coupons, to compare stores, or to deliberate before buying. Imagine, then, the greater stress of purchasing in the real world. ■

Quest for best?

"The perfect is the enemy of the good," said Voltaire, far left, while Henry VIII asserted that the English king "never had any superior but God."

The takeaway

Some choice is good. Too much is bad. It causes psychological pain—so much so that even the pleasure of making the best choices may not negate the stress, expenditure of mental and physical energy, and potential regret of decision-making. As you cannot avoid making choices in your daily life, one potential way to increase your happiness is to simplify the process. Strategize the choosing, not just the choice.

Aoccdrnig to a rscheearch sudty at Cmabrigde Uinervtisy, it deosn't mttaer in waht oredr the ltteers in a wrod are; the olny iprmoetnt tihng is taht the frist and lsat ltteer be in the rghit pclae. The rset can be a toatl mses, and you can sitll raed it wouthit a porbelm. Tihs is bcuseae the huamn mnid deos not raed ervey lteter by istlef, but the wrod as a wlohe.

How we read

Experiment

2.7

Read

the words on the facing page.

Now turn the page.

What happened

The text you just read is a hoax. A compelling, scalp-scratching hoax, but a hoax just the same.

This scrambled paragraph circulated on the Internet in 2003. When it came to the attention of researchers at Cambridge University, they denounced the fraud. Then they explained how, while it contained a grain of truth, it also flew against linguistic research.

First, the truth. Shape is one variable, out of many, that the brain uses to decode printed words. James Cattell, the founder of psycholinguistics, proposed the word-shape model of reading in 1886. He flashed words and letters at test subjects for fractions of a second to test memory. Surprisingly, they recognized words, and the letters they contained, better than letters alone. Cattell concluded that their brains processed word shapes all at once. For example, the brain would recognize the word *shape* by the ascending *h* in the second position and the descending *p* in the fourth.

Matt Davis of Cambridge's Cognition and Brain Sciences Unit has noted that the claim about the significance of the first and last letters is false. He provides this sentence, in which words' first and last letters are intact, as a test: "A dootcr has aimttded the magltheuansr of a tageene ceacnr pintaet who deid aetfr a hatospil durg blendur." Unscrambled, it says, "A doctor has admitted the manslaughter of a teenage cancer patient who died after a hospital drug blunder." The hoax text contains short words that were never scrambled and many words that can be guessed from context. Davis believes letter order, word shape, and context all play roles in reading. ■

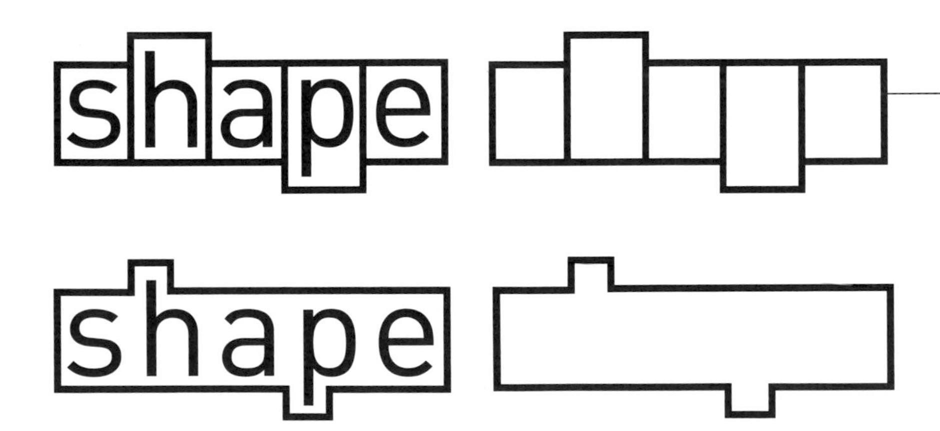

Word contours

Your brain begins to amass a storehouse of printed words by learning letters and then linking them to create meaning. But as you read faster, you seek clues such as shape and letter order to identify words on a line of type.

Look closer

It takes considerable mental effort for a beginning reader to unlock the meaning of black strokes on a white page.

The classic method is to teach a child the names and sounds of individual letters in the alphabet—the ABCs—and then show how letters join hands to make words.

You might assume, then, that people read letter by letter. But experts have debunked the so-called serial letter recognition model of reading. According to this model, the brain processes each letter of a word, starting with the first. Like looking up a word in the dictionary, the brain supposedly goes from letter to letter until it builds an entire word and then recognizes it. The model does have some evidence in its favor, as the brain's linguistic circuits register the meanings of short words more quickly than the meanings of long ones. However, Cattell's word-shape model, also called the word superiority effect, undermines this theory. As Cattell demonstrated, the brain picks out letters more quickly in the middle of a word than in isolation. If the serial letter recognition model were valid, a letter in the third position of a seven-letter word should take three times as long to identify as a letter standing by itself, and this is not the case.

Connections

You do not read letter by letter. Instead, parts of a word link together in the visual and reading circuits of your brain.

Word blindness

Pringle Morgan of Sussex, England, examined a 14-year-old boy in 1896 and declared he suffered from innate "word blindness."

Percy F. had "always been a bright and intelligent boy, quick at games, and in no way inferior to others of his age," Morgan wrote in the *British Medical Journal*. Percy's problem was that he never learned to read. Nor could he.

"The inability is so remarkable," Morgan wrote, "that I have no doubt it is due to some congenital defect." Competent teachers long had struggled to get Percy to master the simplest writing. A bit of Percy's writing appeared this way: "Now you word me wale I spin if. Calfuly winder the sturng rond the Pag." (Now you watch me while I spin it. Carefully winding the string around the peg.)

Morgan's article contained the first scientific description of what now is called dyslexia.

Doctors in the 1920s prescribed eye training for dyslexics. Later research shifted attention to the neural circuitry related to the processing of language. The brain appears to be hardwired for spoken language, but not for reading, which is a relatively new human development. A child learning to read must exert considerable mental effort to understand how sounds are encoded in letters and letters decoded as sounds.

Scans using fMRI have shown that while powerful readers read, strong activation occurs at the rear of the brain in areas including the angular gyrus, home to decoding of print, and Wernicke's area, which processes meaning. Dyslexics display weaker activation in these places but stronger signals in Broca's area, as if their brains were trying to compensate.

Letters all at once

A more widely accepted theory of reading is called parallel letter recognition. A simplified explanation of this complex model holds that the brain grasps individual letters of a word simultaneously. It then compares them with a sort of lexicological database, built by experience, to arrive at meaning. If the word is unfamiliar, children (and adults) read it in pronounceable chunks, not letter by letter, and then compare its parts with those of more familiar words.

Computers that track eye movements provide evidence supporting the parallel recognition model. When you read, your eyes don't move smoothly like a car coasting along a road. Instead, your eyes make tiny jumps, called saccades, of about seven to nine letters, like a car stuck in stop-and-go rush-hour traffic. Unlike the car, the fovea also jumps backward from time to time, although you're not consciously aware of it. The fovea, the tiny spot

> Children have to be educated, but they have also to be left to educate themselves. Abbé Ernest Dimnet, *The Art of Thinking*

Saccades

In 1879, French professor Émile Javal observed that, contrary to common belief, a reader's eyes do not glide across a page, as shown in this diagram of a Swedish text. They jump. They stop. They double back. Each jump takes in information.

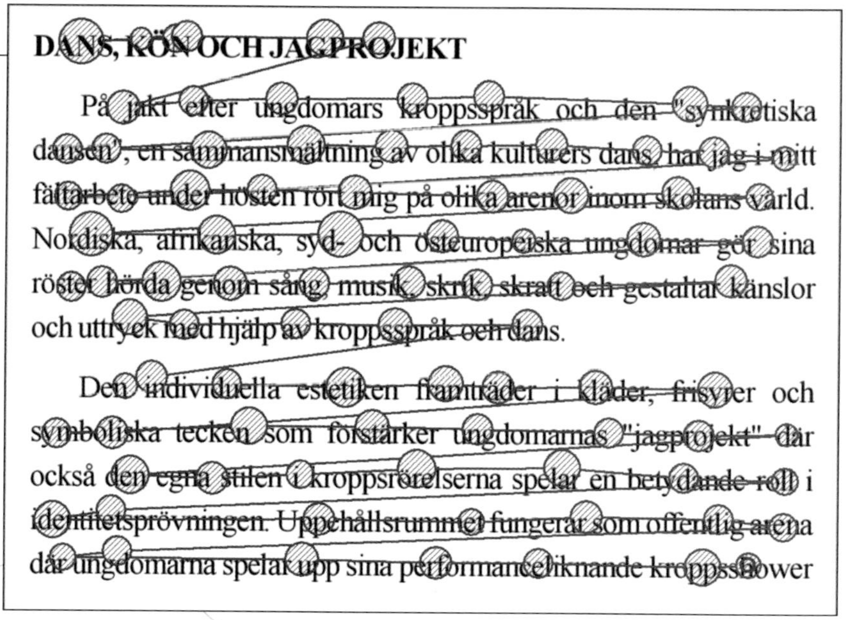

DANS, KÖN OCH JAGPROJEKT

På jakt efter ungdomars kroppsspråk och den "synkretiska dansen", en sammansmältning av olika kulturers dans, har jag i mitt fältarbete under hösten rört mig på olika arenor inom skolans värld. Nordiska, afrikanska, syd- och östeuropeiska ungdomar gör sina röster hörda genom sång, musik, skrik, skratt och gestaltar känslor och uttryck med hjälp av kroppsspråk och dans.

Den individuella estetiken framträder i kläder, frisyrer och symboliska tecken som förstärker ungdomarnas "jagprojekt" där också den egna stilen i kroppsrörelserna spelar en betydande roll i identitetsprövningen. Uppehållsrummet fungerar som offentlig arena där ungdomarna spelar upp sina performanceliknande kroppsshower

of greatest retinal acuity, usually lands on a point just to one side of the center of a word and brings a portion of the word into sharp focus—perhaps three or four letters to either side of the concentration point. The surrounding regions fall under softer resolution.

Researchers have unearthed evidence for this theory by manipulating texts being read in real time. Subjects begin by reading normal words at the start of a sentence. They are unaware of garbled words placed in the text, just ahead of their next saccade. As their eyes begin to make the jump that would discover the error, a camera registers eye movement and substitutes the correctly spelled word without the subject detecting the switch. ■

* The takeaway

Reading is a complex activity integrating unconscious motor skills to move eyes in discrete jumps, sensory skills to recognize words by their letters, and cognitive skills to decode meaning. How complicated, yet how sublime, that reading creates sounds and pictures in the mind. A computer with optical software may scan Joyce Kilmer's "Trees," but only the human brain can read it and hear the rustle of the leaves and picture their shapes and colors.

Chapter 3

BE

I N G

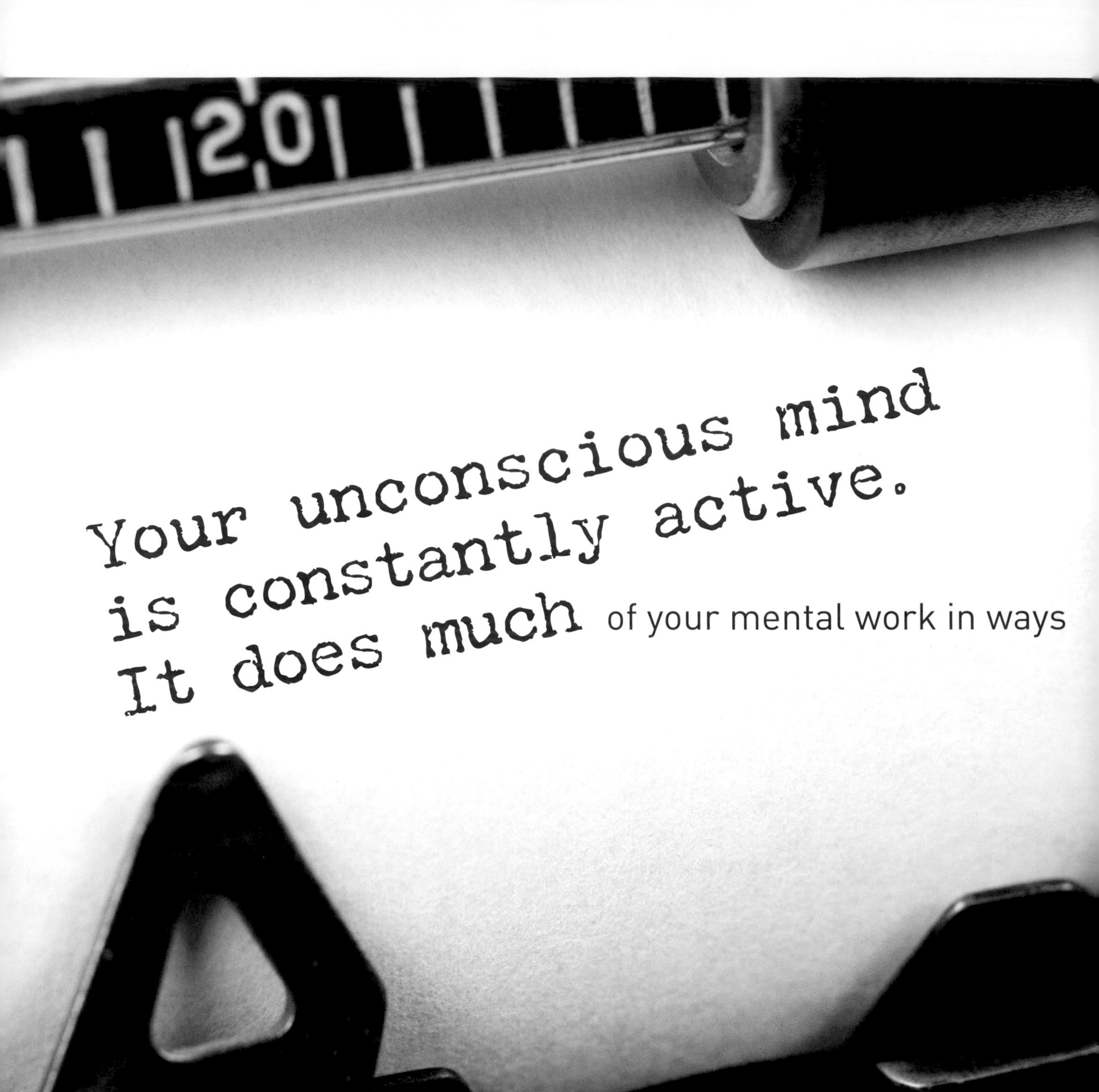

Your unconscious mind is constantly active. It does much of your mental work in ways

Illusions commend themselves to us because they save us pain and allow us to enjoy pleasure instead. We must therefore accept it without complaint when they sometimes collide with a bit of reality against which they are dashed to pieces. Sigmund Freud

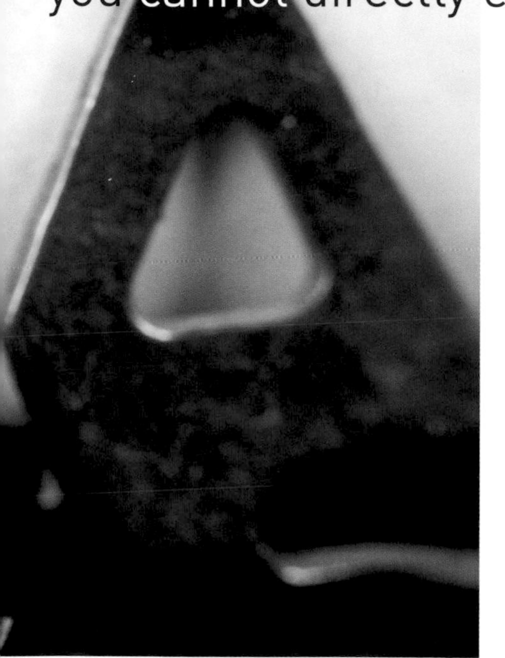

you cannot directly experience. Many actions you take occur not because you consciously choose them, but rather because unconscious processes led you to them. • Your brain makes untold calculations to maintain control of a car on a curving highway in the rain or to swing a tennis racket to intercept a drop shot before it hits the ground. Before you get too proud of such feats, consider that unconscious processes also guide bees to pollen-bearing flowers, salmon to their native streams, and migrating birds to distant countries. • And yet humans dismiss the role of the unconscious mind as somehow secondary to the conscious functions occupying the executive regions of the frontal lobes. Humanity, after all, is *Homo sapiens*—"wise man"—and we are quick to emphasize that wisdom, conscious wisdom, makes our species unique.

The center that

Mere consciousness

To appreciate the role of your unconscious mind, imagine life without it. Psychologist Timothy D. Wilson poses just this scenario in his book *Strangers to Ourselves: Discovering the Adaptive Unconscious*. Suppose a man awoke one day and, unbeknownst to him, his unconscious mind had shut down during the night. How would his life differ? Seventeenth-century French philosopher René Descartes would have expected no change. He recognized no unconscious mind; instead, he defined the human animal by its awareness of its own thoughts. "Cogito, ergo sum," he said—"I think, therefore I am."

Wilson disagrees. A person who loses unconscious abilities would not even be able to get out of bed. The unconscious mind automatically monitors the position of the body and the locations of its many parts. It orders muscles to contract at just the right times, in just the right ways, to maintain the body's balance. Without this unconscious process, the purely conscious man would fall and not get up.

That same man also would be overwhelmed with sensory stimuli. Bathed in millions of bits of information striking the nerve endings of his five senses, he would struggle to make order of the chaos. There would be no resolution of the colors and lines of his bedroom into a coherent, three-dimensional image through which he could navigate.

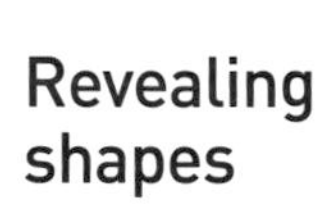

Revealing shapes

An inkblot test inspired by the psychological Rorschach exam was thought to allow assessment of emotional functioning. Such tests generally have been discredited.

Unconscious understanding

If the man with no unconscious mind managed to crawl to the television and turn it on, he would wrestle with the sentences voiced by the actors, newscasters, and advertisers. Their words would challenge his understanding because the unconscious mind quickly assigns particular meanings to words and phrases by automatically checking definitions, associations, contexts, and so on. For example, if a character says, "That's amazing," the purely conscious man would have no idea whether the speaker felt awe or was being sarcastic.

If the man tried to converse with friends or neighbors, it's doubtful he would speak with anything approaching eloquence. Speech relies heavily on unconscious processing of language. When you converse, you don't consciously map the grid of every sentence before vocalizing it in order to make it grammatically correct. Your conscious mind, which works more slowly than its unconscious cousin, doesn't have time to do so and still keep up with conversation.

Mental bureaucracy

"The mental processes that operate our perceptual, language, and motor systems operate largely outside of awareness, much like the vast workings of the federal government that go on out of view of the president," Wilson wrote. "If all the lower-level members of the executive branch were to take the day off, very little governmental work would get done."

Left versus right

Although the right and left halves of your brain share some functions, not all functions are divided evenly. On the left usually are the centers for understanding and using language, including reading and writing, and manipulating details. On the right usually are the centers for spatial awareness and information integration.

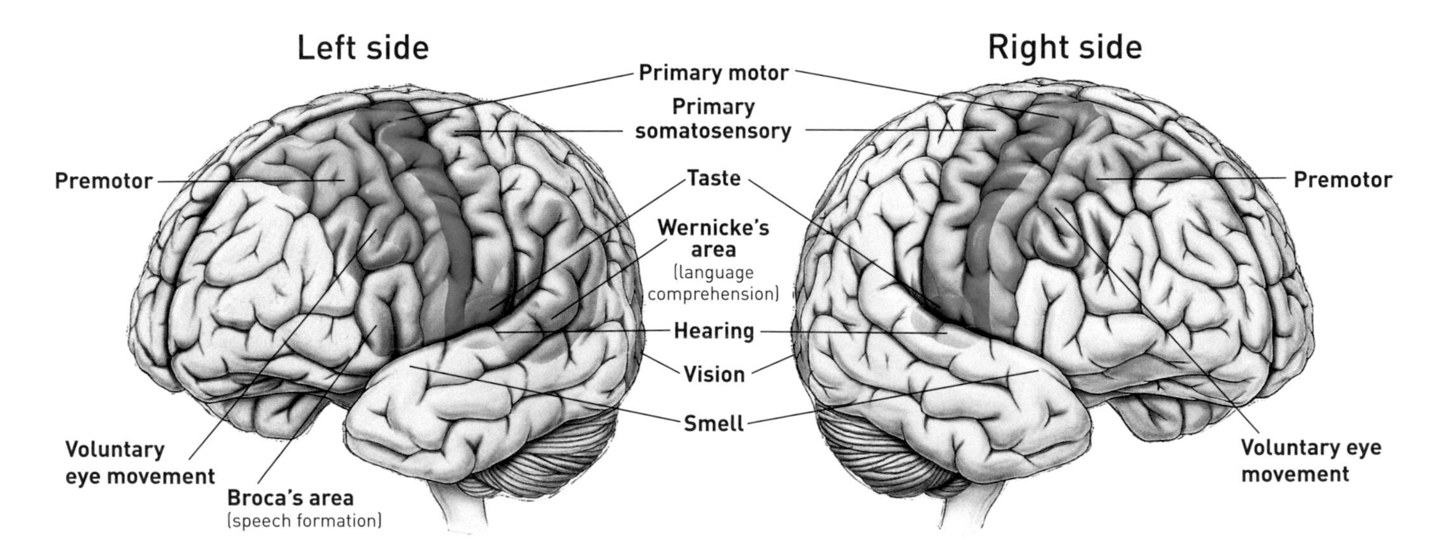

Without thought

Once you've mastered riding a bike, you don't need to think about it—at least, not consciously. Your unconscious mind constantly monitors your body position and orders adjustments to your balance so you don't fall.

To extend the analogy, the upper levels of government collect data from thousands of civil servants and then decide what to do with it. Behind the executive function of the conscious mind, the unconscious continues to play a hidden role—the power behind the executive desk. It relies on memory to script the mind and body to act quickly in familiar situations, such as crossing a busy street, avoiding a snarling dog, or nodding hello to an acquaintance in the hall. It prompts habitual behaviors in social situations and thus creates much of your personality.

Even decisions we experience as conscious, such as choosing a meal, a car, or even a mate, have been quietly led through shortcuts by our unconscious mind. For example, the unconscious mind prejudices behaviors that appear to be under our control, even when control is impossible. It changes the value people place on objects once they fall into their possession, thus prejudicing behavior in favor of retention of resources. And it shapes the way people remember pleasure and pain, factors that influence future decisions.

The emotional unconscious

In the last three decades researchers have begun making the case for the existence of a powerful, *emotional* unconscious. In the 1980s, social psychologist Robert Zajonc demonstrated that people can form preferences without any conscious awareness of sensory stimuli. Zajonc argued that because preferences are emotional reactions—things prompt feelings of happiness or sadness, for example, so you favor or disfavor them—emotion can exist before, or without, cognition.

Zajonc and his colleagues conducted experiments in which they showed test subjects some unusual visual patterns, including Chinese ideograms. Later, when the same subjects observed new ideograms along with some of the ones they had seen before, they expressed preference for the familiar ones. Simple exposure created the preference. Zajonc later decreased image exposure to such a short time that the subjects had only a subliminal experience. The subjects could not say what had been flashed before their eyes because the images did not exist long enough to register on conscious thought. Nonetheless, they still preferred the familiar (if unknown) images.

Subliminal priming

Zajonc also found evidence for subliminal emotional priming. He showed pictures of frowning or smiling faces to his experimental subjects for about five milliseconds and then

The iceberg's tip

Your conscious mind constitutes a fraction of the work of your brain. Without the nonconscious control of tasks such as regulating heartbeat and respiration, moving muscles, processing emotions, and so on, your conscious mind would bog down in the demands of simply living.

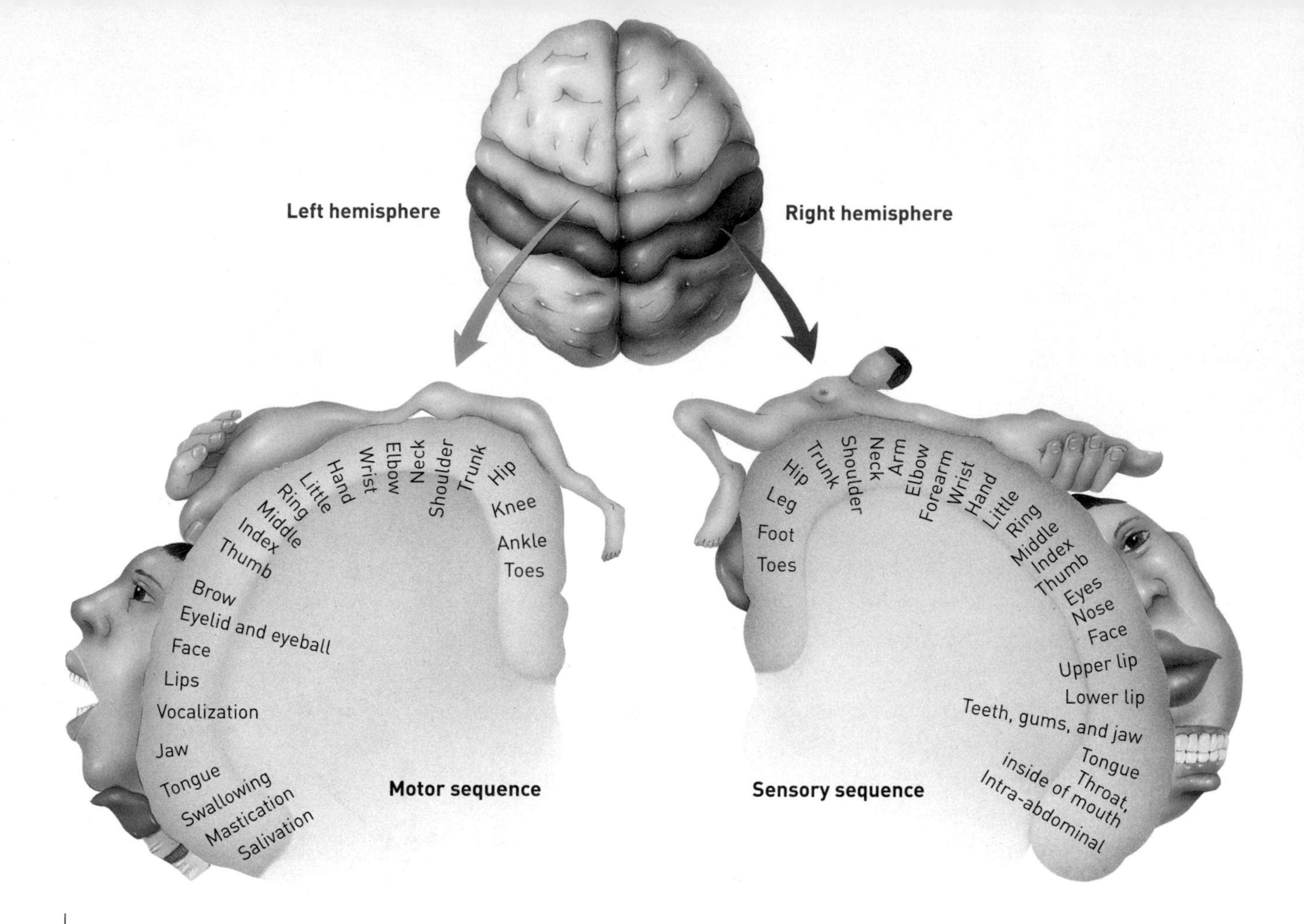

Body maps

The motor homunculus (at left) and somatic sensory homunculus (at right) exaggerate body parts to illustrate the relative amount and location of neurons in the elevated areas of the cortex devoted to movement and sensation. The primary motor cortex is marked in blue, the somatic sensory cortex in purple.

followed the image with a masking image. The short exposure followed by the mask prevented them from recalling what they had seen. After each preliminary exposure there appeared an emotionally neutral second image, such as a Chinese ideogram, that stayed visible long enough to be consciously perceived. When Zajonc asked the subjects whether they liked or disliked the second image, he found a strong correlation. Liked images followed smiley faces. Disliked images followed frowny faces. Subliminally, your unconscious mind pays attention to things that your conscious mind cannot—a fact that advertisers use to their advantage.

Psychologists don't just describe thoughts in terms of conscious and unconscious. They also categorize them as fast and automatic or slow and reflective. The fast and automatic ones can create illusions of perception and memory, as seen in Chapters 1 and 2. But slow, reflective

> The unconscious mind of man sees correctly even when conscious reason is blind and impotent. Carl Jung

thoughts also can lead to illusions. As you deliberate such things as how you might feel if your best friend died, or why you insist on a misdeal when cards get dealt out of turn, you likely fall into mental traps.

Fast, unconscious processing of information, from sensation to emotion, lays the foundation for the brain's higher and slower functions. "It is probably not too far off the mark to say that consciousness will only be understood by understanding the unconscious processes that make it possible," said neuroscientist Joseph LeDoux.

Intuition versus reason

The mind needs both unconscious intuition and reasoned judgment. Intuition reaches quick conclusions about things being good or bad. It's a fast and easy pathway to decisions. Many intuitive decisions are also logically sound because the brain accesses emotional memories from similar situations in the past and applies them to a snap decision in the present.

Intuition based on emotional responses can be a powerful thing. Wilson and colleague Jonathan Schooler asked test subjects to do a blind test and rate the tastes of five brands of strawberry jam. One group then wrote a short essay about their college major—a cognitively neutral task—before rating the jams on a numerical scale. They never tried to articulate reasons for liking some more than others. Another group wrote reasons for liking or disliking each jam and then rated the jams on the same scale. Wilson and Schooler compared the two groups' jam ratings with those of *Consumer Reports* magazine. Strangely, the students who wrote their reasons for liking or disliking particular jams had almost no correlation with the jam ratings of *Consumer Reports.* Meanwhile, the students who wrote about something totally unrelated to the taste test—their college majors—exhibited ratings significantly more in line with those of the expert tasters. The results seemed to suggest that thinking hard made the ratings worse.

According to psychologists Christopher Chabris and Daniel Simons, who described the jam study in their book *The Invisible Gorilla,* two

Uncertainty

If you think you know how you think, you're probably wrong. Humans place a great deal of confidence in reason, a quality they alone possess in the animal kingdom. Yet emotion, intuition, and other mental functions quietly support good decision-making.

Fact

A fifth of your blood is devoted to working your brain. Your brain needs this disproportionate share of blood to meet the metabolic demands of its billions of active neurons.

reasons explain this oddity. First, conscious thought adds nothing to the knowledge of jam. And second, preferences for particular tastes aren't formed intellectually. They arise at some baser yet faster emotional level. Thinking just jams up thoughts with irrelevant information.

Nevertheless, slow and reflective thoughts obviously help us make some sound decisions—why else would natural selection favor the power of reason? According to Chabris and Simons, "Deliberation will outperform intuition when you have conscious access to all the necessary data."

The slow, deliberative mind lets people remember situations in the past, recall how their decisions might have led to good or bad outcomes, and then project how currently contemplated choices might lead to future outcomes. Consciousness lets the brain imagine and plan.

Imagination and plans

When the ability to plan fails, something fundamental goes out of the mind. A patient known as N.N. lost his ability to make plans when he suffered damage to his frontal lobe, the seat of executive functions, in a 1981 car accident. Asked to imagine what he would do the next day, N.N. struggled. When a psychologist asked a follow-up—to have N.N. describe his state of mind when he thought about the future—the patient responded, "Blank, I guess . . . It's like being asleep . . . Like being in a room with nothing there and having a guy tell you to go find a chair, and there's nothing there . . . Like swimming in the middle of a lake." And yet, N.N. could carry on intelligent conversations. His injury did not reduce his knowledge of the world; it only took his ability to imagine how he might interact with that world.

N.N. lived in a permanent present. Such a description matches what scientists now believe describes the state of the human brain early in its evolution. Only when their frontal lobes developed, bringing them an awareness of their world, past, present, and future, did humans become recognizable as modern people. To think, imagine, and plan is to be human. To err also is human. And that is no coincidence. ■

Brain thoughts

In the words of Woody Allen (opposite), the brain is "the most overrated organ." Still, perhaps no subject has been the target of so much intellectual firepower as the mind itself and its central role in the human experience.

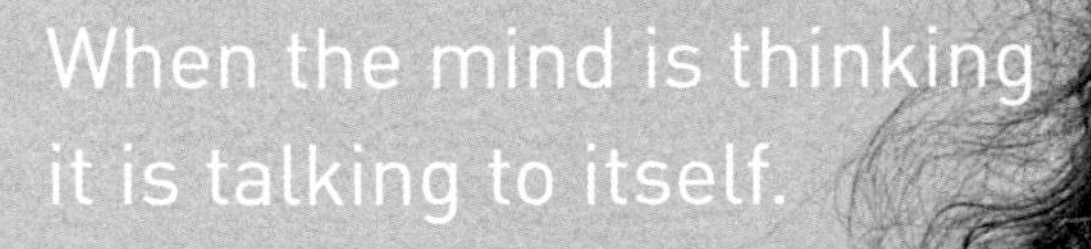

When the mind is thinking
it is talking to itself.

Plato

An intellectual is someone
whose mind watches itself.

Albert Camus

Our reason is always
disappointed by the
inconstancy of appearances.

Blaise Pascal

Man consists of two
parts, his mind and his body,
only the body has more fun.

Woody Allen

65

$1.00

65

Illusion of control

Experiment 3.1

Imagine that your

office has a lottery. All 200 tickets sell for one dollar apiece.
At the end of two weeks, a drawing will determine the $200 winner.

You buy a ticket

and choose the number 65, for the day that you got married, June 5.

Now imagine that

a co-worker asks you to trade tickets because she likes
that number better than her 23. Would you trade, even-steven?
If not, what would you consider a fair deal?

Now turn the page.

What happened

If you're like most people, you'd be reluctant to trade your ticket for another one, although each has exactly 1 chance in 200 of winning.

By selecting the number and purchasing your ticket, you likely consider it "your" ticket. Furthermore, your freedom of choice endowed the selection with what Harvard psychologist Ellen Langer calls "the illusion of control." It reflects the brain's preference for controlling situations, even the uncontrollable ones. It explains why some people blow on dice before a toss, or use a strong motion when trying to throw high numbers but a weak motion for low ones.

A natural roll?

Do you blow on dice to roll a seven? If so, you're demonstrating mental bias for control, even when control is impossible. Controlling actions may occur at or below the level of consciousness.

Langer's research agenda sprang from a poker misdeal that she witnessed as a graduate student. The dealer skipped a bettor while distributing a round of cards. "Everyone went crazy," Langer said. Nobody ever considered the simplest solution: to have the dealer assign the next card to the person who was one short. Despite the randomness of a shuffled deck, the players considered "their" cards to arrive in a particular order.

Psychology at the time focused on behavior as rational. But clearly, gamblers made irrational choices based on the idea of having some control over luck.

To test this idea, Langer ran an experimental lottery. In one group, participants paid a dollar and chose a football player card as their lottery ticket. In the other group, the experimenters assigned everyone a football card in return for a dollar. Players then stated what price they would demand for their ticket. The ones who made choices asked an average of $8.67. The others wanted only $1.96. ■

Ellen Langer's home page: **http://www.ellenlanger.com/home/**
Examines decision-making, the illusion of control, and her advocacy of "mindfulness."

Look closer

Much human behavior occurs outside the realm of rational, conscious thought.

Some researchers, such as psychologist Daniel Gilbert and neurologist Antonio Damasio, take that conclusion to heart. They look at decisions that the brain makes while seemingly on autopilot. They have found the unconscious mind, thanks to millions of years of evolution that preceded the emergence of consciousness, doing a pretty good job of navigating the body through potential dangers.

Many behaviors occurring outside consciousness, such as instincts, are extremely important. When you hear the buzz of a rattlesnake's tail or the snarl of a large dog, your unconscious mind charges your body to fight or take flight before your conscious awareness registers the threat. In addition, quick emotional reactions often prove to be as good as or better than ones reached through deliberation.

Other researchers, such as Langer, focus on the pitfalls of automatic behaviors, such as stereotyping. She advocates "mindfulness" over "mindlessness" to better appreciate life.

Mind over matter

Langer's research has blossomed into demonstrating the power of mind over matter. She found that after telling one group of hotel workers that their daily work constituted exercise, on par with being in a gym, within two months these workers lost an average of two pounds and lowered their blood pressure ten points despite maintaining the same diet and work habits. This group was healthier than another group of workers, who were doing the same work but were not "mindful" of its beneficial effects. In another experiment, men in their 70s and 80s exhibited better psychological and physical health after being asked to relive their youth in a setting designed to re-create 1959.

And, in an experiment that echoed the illusion of control,

Alert! Alert!

Your unconscious mind monitors the world and prompts reactions that boost your chances for survival. You don't need to think to react to the sound of an angry rattler's tail.

CASE STUDY

The invisible violinist

How much can the conscious mind process at once? Or, to put it another way, how prevalent is the "mindlessness" studied by Langer?

The brain processes about 11 million bits of sensory data every moment. Researchers arrived at that number by counting receptor cells in the five senses and their associated nerve pathways. And yet, the most generous guess is that the conscious mind can process only about 40 bits at a time. The rest get processed unconsciously.

That's generally a good thing. Being aware of everything would overwhelm you. By screening out all you consider extraneous, you can focus on what's important to you. But in a phenomenon known as inattentional blindness, you miss a lot right in front of your nose. At Harvard, psychologists Christopher Chabris and Daniel Simons conducted a famous experiment in which about half the subjects who scrutinized a short video clip failed to see a woman in a gorilla suit walking through the frame. Tracking sensors revealed that those who didn't see the gorilla had their eyes land squarely on it, but they failed to register it. They had been told to count the basketball passes in the video, so that's what they did.

Could such results be duplicated outside the lab?

The *Washington Post* won a Pulitzer Prize in 2008 for a real-world equivalent. During rush hour, writer Gene Weingarten took virtuoso violinist Joshua Bell and his Stradivarius to the crowded L'Enfant Plaza subway platform in Washington, D.C. Bell played complex classical pieces for 43 minutes. More than a thousand people lost in their own thoughts passed the once-in-a-lifetime concert; only seven paused to listen. Bell earned $32.17 in tips. He said afterward that he was "surprised at the number who don't pay attention at all, as if I'm invisible."

Langer documented improved well-being in nursing-home residents given a greater say over their lives. She gave one group a houseplant and said they were in charge of the plant's feeding and watering. A second group also received a plant but were told the staff would take care of it. Within six months, 15 percent of the group given control of the plant had died, compared with 30 percent of those without the responsibility. A similar study found a significant health difference between those who controlled the timing and duration of visits by student volunteers and those who had no such control.

Sports rituals

Superfans can't really influence their teams' chances of victory. Yet fans often repeat rituals, such as sitting or acting in ways that correlated with previous athletic success. Lucky hat, anyone?

The joy of control

"The fact is that human beings come into the world with a passion for control, they go out of the world the same way, and research suggests that if they lose the ability to control things at any point between their entrance and their exit, they become unhappy, helpless, hopeless, and depressed," Gilbert said.

Desire for control grips the mind so tightly that it believes it can conquer random chance. This creates a feeling of reward. According to Gilbert, the forebrain desires control because it perceives making choices as the path toward an imagined, better future. This quality makes humans unique among all organisms; only people can imagine a future beyond this moment and seek ways to make it a happy one. ■

* The takeaway

Your brain wants to be in charge of actions affecting your future. This feeling is so strong, and so rewarding, that you likely experience the "illusion of control" and act in some illogical ways. If you think sitting in a particular seat or doing a certain cheer will help your favorite sports team win, or if you insist on a particular ritual in a game of chance, you are only being human.

Illusion of intention

Experiment 3.2

Imagine

you're in a college acting class. A professor assigns you to prepare a dramatic speech as if you were someone famous. To challenge yourself, you choose someone who's not a tyrant or a monster, but rather a political figure you dislike. The big day comes, and you give a great speech.

Now, ask yourself:

Have you changed your attitude about the person you mimicked?

Now turn the page.

What happened

Your appreciation of the famous figure likely would grow. Scientists have replicated versions of this experiment many times and found that to be true.

But that's not all. After giving the speech, you also might rationalize the choice by unconsciously altering memories of your old attitudes.

Like Aesop's fox who wanted the grapes but pronounced them sour after he could not reach them, your brain plays with your intentions after you act.

In proposing his theory of cognitive dissonance in 1957, psychologist Leon Festinger argued that people revise feelings to be consistent with actions. Any inconsistency, such as your giving a well-received speech in the role of a politician you dislike, causes mental discomfort. The brain resolves the conflict by changing attitudes to match behaviors. If you buy a car with the knowledge that it has a bad transmission, you might actually come to value the car *more* because of the defect, as if you had decided it must be terrific despite the flaw. Neural circuitry that carries out action is distinct from circuitry that explains action. People not only can, but often do, act unconsciously. Humans feel a need to justify their actions because of their perception of conscious will. Sometimes they invent intentions out of whole cloth.

In the case of cognitive dissonance, "the theory suggests that actions can 'sneak by' without sufficient intention," wrote Harvard psychologist Daniel M. Wegner, "but that once having sneaked, they can become unpleasantly inconsistent with those contrary prior intentions and prompt the revision that creates a new intention." ■

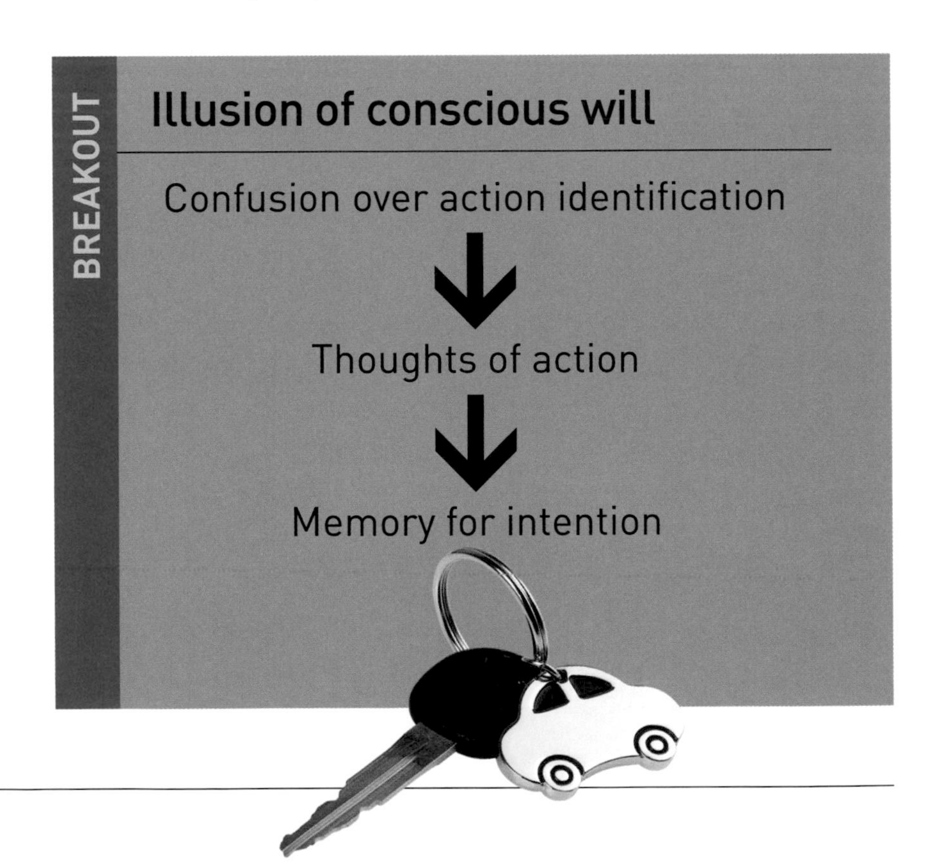

Look closer

Sigmund Freud, founder of psychoanalysis, knew that most mental activity occurs unconsciously.

He uncovered buried drives and emotions underlying odd behavior, but there is a more innocent side to unconscious action. How many times have you walked into a room and totally forgotten why you entered it? Or started talking to yourself without any expectation of doing so?

These types of behaviors occur because the brain segregates action from explanation. In his book *The Illusion of Conscious Will*, Wegner lists three sources for unconscious action that challenge the notion of a conscious mind totally controlling behavior: confusion over action identification, thoughts of action, and memory for intention. Unconscious actions, like cognitive dissonance, often occur as post-event revisions.

Sources of confusion

Confusion arises about identifying actions because every behavior has a variety of possible meanings. Lying on the beach could be a sign of getting a tan, feeling tired, being on vacation, feeling sick, or examining seashells at close range.

People tend to assign only one meaning per action. Because of the brain's flexibility of choice, however, you might give one meaning while the action is under way and change it later. Someone shooting an intruder might cast it as self-defense but revise it as an accident.

Confusion over thoughts of action includes the role of unconscious thoughts in everyday situations. Freud plumbed these depths when he postulated the impact of repressed sexuality. More modern examples include unconscious priming to think certain things. People with phobias about snakes think about snakes a lot. Likewise, priming thoughts about stereotypes evokes evidence of stereotypes in everyday situations.

Confusion about memory for intention stems from the fallibility of memory itself. Memories fade and change. The brain has a memory system to recall intentions for actions in the past. It too

Ouija boards

A paradox of consciousness is that by definition, the conscious mind cannot directly examine unconscious actions—if it could, those actions would be conscious ones. You don't know what you can't know.

However, one way to study unconscious action is to examine its physical effects.

Say hello to exhibit A, the Ouija board.

Toymakers began producing the Ouija board in 1890. Instructions call for two or more players to place their fingertips on a three-legged planchette and allow it to move over a board—seemingly under the volition of an unseen spirit—as they ask questions. The planchette often glides from letter to letter to spell words.

Science provides two explanations for planchette movement. Either the operators know they move it but say nothing, or they move it unconsciously.

Evidence for the latter conclusion comes from the Victorian-era phenomenon of table turning. Hoping to contact the spirit world, participants sat at a table, rested their hands atop it, and waited. Tables often spun and rose in the air, but nobody claimed responsibility. Michael Faraday, the British chemist, physicist, and electromagnetic pioneer, set out to solve the mystery in 1853. He put stacks of slippery paper between the participants' hands and the tabletop. If the table turned by itself, the bottom of the stacks would slide away from the hands. But if the hands moved the table, they would pull the top of the stacks with them. Faraday's experiment proved that the table turners imparted motion without conscious knowledge.

In groups, small motions sometimes cancel each other or combine dramatically. So it is with Ouija.

is fallible, leaving people open to doing things with no memory, or a false memory, of why.

Inventing explanations

As humans believe they act rationally but often exhibit irrational, unconscious behavior, their brains find creative ways to deal with unconscious actions.

Mental processes of patients who have had split-brain surgery offer a glimpse into how the brain does this. Michael S. Gazzaniga studies people whose left and right cortices have been separated by cutting the corpus callosum, the thick band of fibers that provides the only connection between our left and right hemispheres. Such extreme surgery provides relief for chronic seizures and has surprisingly few side effects. It

Intuition is the clear conception of the whole at once. Johann Kaspar Lavater

Bridging halves

A band of connective tissue called the corpus callosum, hidden in the cleft at right, lets information pass between the brain's hemispheres.

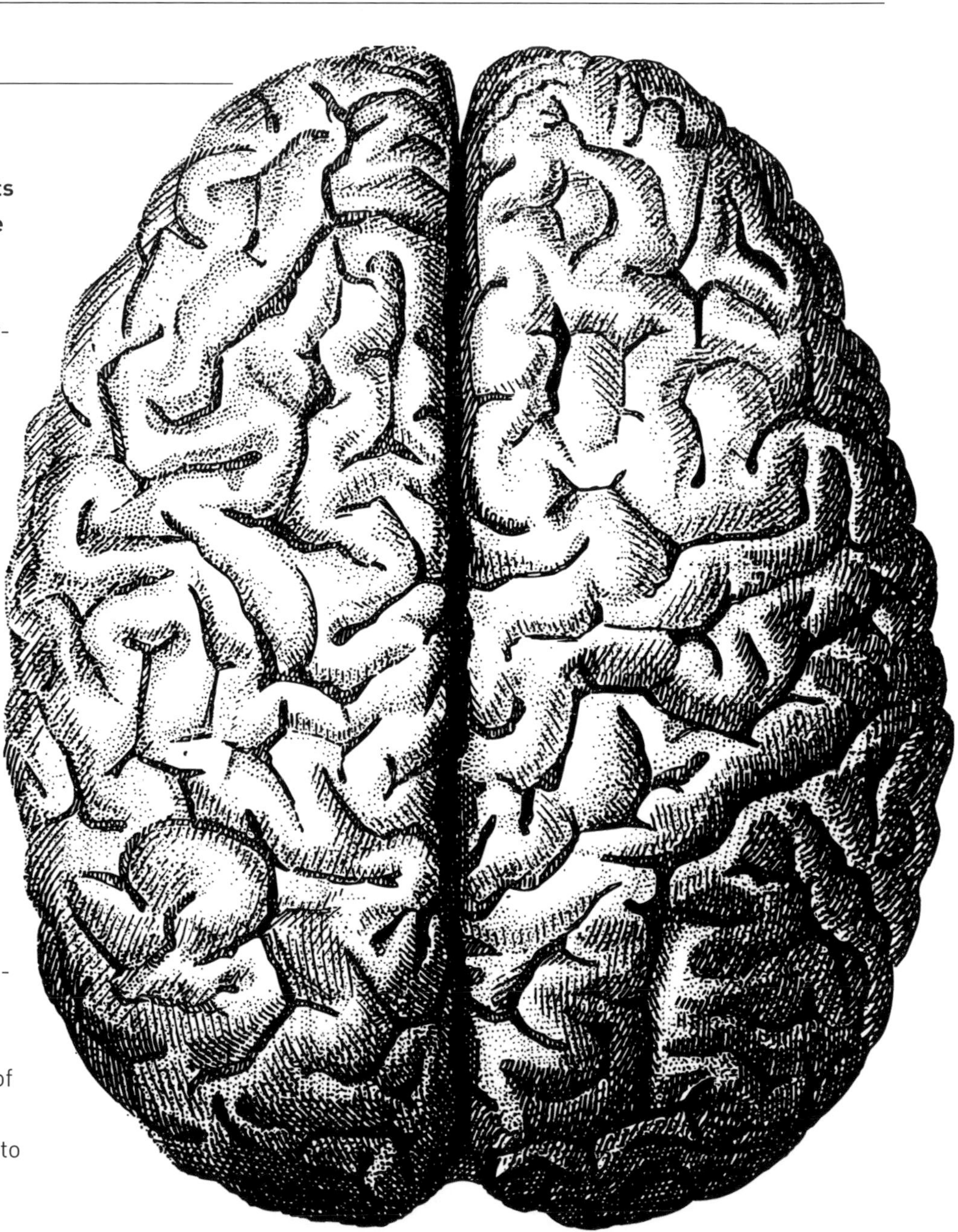

blocks the transmission of information from one hemisphere to the other, but people are able to act normally, despite the independence of each of their isolated hemispheres.

Two hemispheres

Physicians long have known that damage to one side of the brain, such as a soldier's shrapnel wounds, can impair body movement on the other side. A wound to the right hemisphere, for example, could paralyze the left hand. In addition to controlling the body's right side, however, the left hemisphere is also the center of verbal activity.

Examinations of split-brain patients who possess a shred of verbal skills in the right hemisphere have yielded insights into the illusion of intention.

PURE WHISKEY

> The [left-brain] interpreter seeks patterns, order, and causal relationships. Michael Gazzaniga

Gazzaniga studied just such a split-brain patient, identified as J.W. Gazzaniga set up a visual experiment in which he flashed separate images to J.W.'s left and right visual fields. When J.W. saw the word *laugh* in his left field, and his right hemisphere processed it, he laughed.

But when Gazzaniga asked why J.W. was laughing, the verbal response arose in the left hemisphere, which had no memory of seeing the word. "You guys come up and test us every month," J.W. said. "What a way to make a living." Gazzaniga concluded that some part of J.W.'s left hemisphere—what he calls the "left brain interpreter"—improvised rational intentions after the fact.

Improbable tales

Historical evidence abounds for the invention of sometimes impossible intentions under other unconscious conditions. In the 19th century, German psychiatrist Albert Moll wrote about people executing post-hypnotic suggestions.

"I tell a hypnotized subject that when he wakes he is to take a flower-pot from the window, wrap it in a cloth, put it on the sofa, and bow to it three times," Moll wrote. "All which he does. When he is asked for his reasons, he answers, 'You know, when I woke and saw the flower-pot there I thought that as it was rather cold the flower-pot had better be warmed a little, or else the plant would die. So I wrapped it in the cloth, and then I thought that as the sofa was near the fire I would put the flower-pot on it; and I bowed because I was pleased with myself for having such a bright idea.'"

Interpretive functions

Gazzaniga's theories about the left hemisphere's function of monitoring and interpreting behavior explain the post-hypnotic story. The interpretive function doesn't need to know why behavior occurs. It merely monitors behavior and then gives it purpose. ■

Hypnosis

Do you will your own actions? Hypnotists, such as one in this early 20th-century poster, appear to bend the will of others.

The takeaway

You may think every action you take or attitude you hold arises from clear and conscious intentions, but that's not true. Thanks to the human desire to cling to the illusion of a rational will, your brain tries to make sense of behaviors with no rational, conscious origin. Your left hemisphere supplies these explanations, sometimes after the fact, and sometimes without your being aware of their implausibility.

PUSH

Unconscious Will

Experiment 3.3

Place your index finger

atop the image of the button.

Count backward

from five, as if you were ticking off seconds until a rocket's liftoff.

When you start to say the word *zero*, mentally order your finger to push the button.

Make your best guess:

At what moment did your mind begin to tell your finger to move?

Now turn the page.

What happened

Inside your brain, electrical activity associated with movement began increasing nearly a full second before your finger jerked at the word *zero*.

Two German researchers, Hans Helmut Kornhuber and Lüder Deecke, demonstrated this gap in a series of experiments in the 1960s. They attached electroencephalographs to the scalps of test subjects to measure brain activity as the subjects executed simple finger movements. The researchers compared the results with recordings of muscle movement detectors called electromyographs. Time after time, they found a precursor spike in the brain 0.8 second before movement. They dubbed it the readiness potential. Sensors showed this generalized activity occurring in the mid-parietal region at the top of the head, and in the left and right motor cortices just above and in front of the ears.

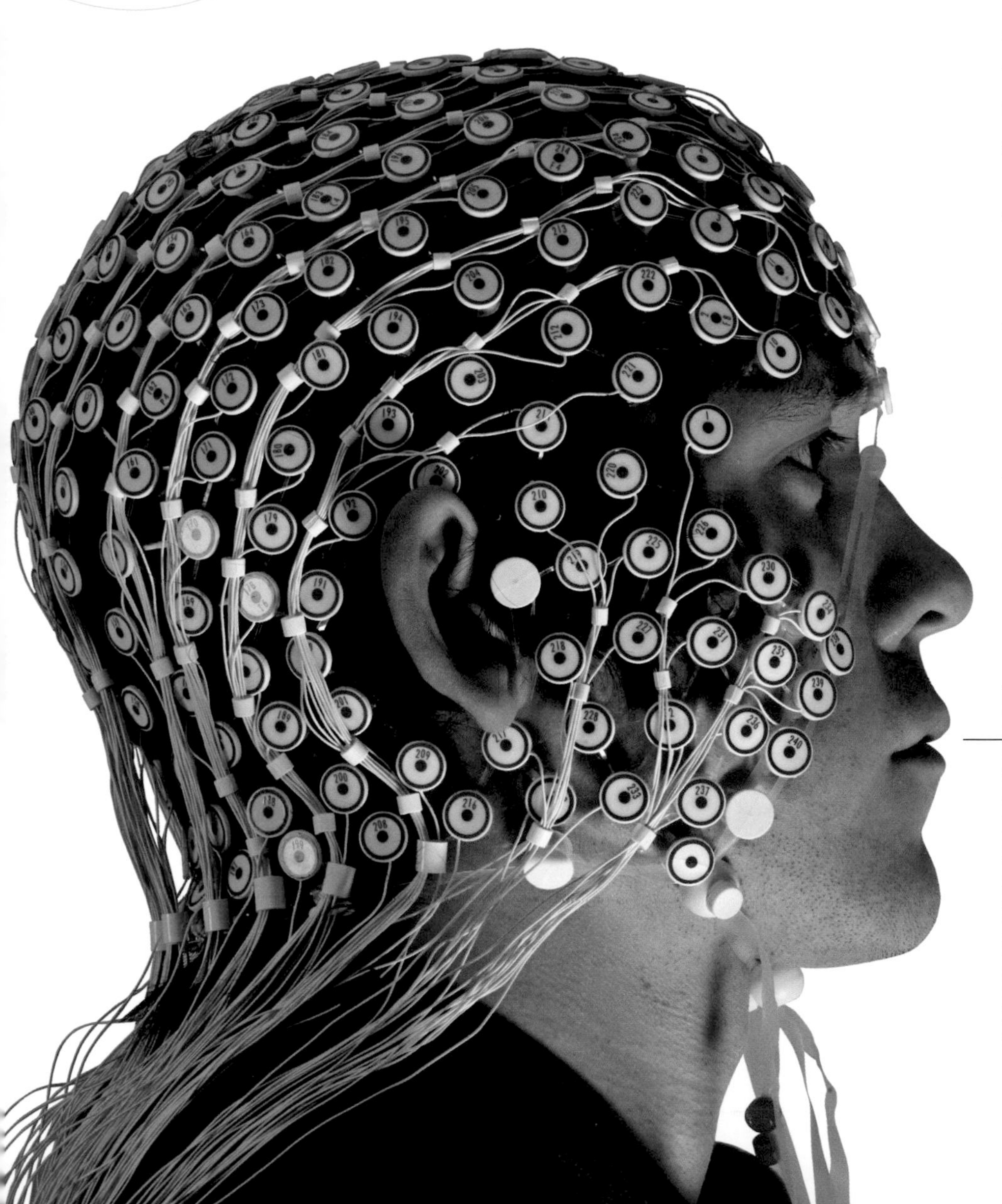

Electric map

A college student has his brain wired for an electroencephalograph to study its inner workings in real time. Precisely timed encephalographs have shown that the brain initiates action before the conscious mind recognizes itself willing the action.

Kornhuber and Deecke also found a second, smaller spark of electrical activity at 0.05 second before finger movement. Unlike the broader readiness potential, it pinged through the specific motor region controlling the fingers. Viewing the two recordings together suggests a generalized activation focusing itself into a specific action.

Kornhuber and Deecke's graphs charted neural firing but not cognition. The latter phenomenon became the quest of another scientist, Benjamin Libet, in the 1980s. Libet also wired sensors to the scalps and fingers of test subjects, but he added a new element: Each volunteer moved a finger at will but used an elaborate clock to pinpoint the moment of conscious awareness of "wanting" to move. Strangely, that moment occurred significantly *after* the moment of readiness potential, but before sensors detected finger movement.

Conclusion: An unconscious part of your brain "wills" an action before you are consciously aware of your will to direct it. ■

! Look closer

The experiments of Kornhuber, Deecke, and Libet suggest that the will of your conscious mind may not be animating "voluntary" acts.

Instead, what you consider willful action may be your brain's attempt to fashion an explanation after the fact. It appears that some part of the brain drives behavior, and another part, which believes it wills the behavior, merely explains it. "You meant to do that," your brain says after a nonconscious force makes you grab a cup, strike a match, or swing a bat at a curve ball.

Libet said, "Clearly, free will or free choice of whether to act now could not be the initiating agent, contrary to one's widely held view."

So much for classical philosophy. For centuries great thinkers have expounded on thoughts causing actions. According to Aristotle, mind sparks desires, which prompt deeds, as when you perceive thirst and drink. French philosopher René Descartes famously stated that mind and body are distinct, but the purely spiritual mind moves

Checked swing?

A fastball heads toward the plate. Do you swing? Part of your brain makes the decision, and another part explains it—and takes credit.

CASE STUDY

Alien hand

When he was 56, a patient identified as J.C. suffered a stroke in his left frontal lobe. It extended into his corpus callosum, the fibers that connect the brain's hemispheres. Four weeks later, his right hand seemed to become possessed. "It has a mind of its own," J.C. told his doctors. "It won't go the way I want."

J.C. tucked his shirt into his pants with his left hand; his right untucked it. He grabbed papers with his left hand; his right ripped them free. He poured tea with his right hand but couldn't stop when the cup was full. He restrained his right hand but couldn't tame it.

His condition, called alien hand syndrome, is very rare—only about 50 cases had been documented between the first in 1909 and J.C.'s in 2006. That didn't stop Hollywood from making it famous. In a memorable scene of 1964's *Dr. Strangelove,* a scientist played by Peter Sellers struggles to keep his arm from giving a Nazi salute.

Alien hand is marked by feelings of foreignness in the affected limb and purposeful movements perceived as involuntary. It springs from damage to the medial frontal lobe, the corpus callosum, or both. The frontal lobe is home to higher-order mental functions including action planning. The right side controls body movement on the left, and vice versa. Damage to the corpus callosum prevents left and right hemispheres from coordinating.

In 2007, Swiss researchers scanned the brain of an alien hand patient. They found planned movements starting in the frontal lobe and then activating the brain's "motor strip." Unplanned movements began in the motor strip without a cue in the frontal lobe. This seemingly spontaneous firing has defied explanation.

the purely physical body as a ghost in the machine. Such arguments would have gained traction if scientists had found evidence of the feeling of conscious will preceding the brain's readiness potential.

Neuroscientist Antonio Damasio, in his book *Descartes' Error,* has taken the French philosopher to task. Damasio, and many others, argue that the notion of the separation of mind and body is wrong. A wealth of evidence now suggests that consciousness arises as an emergent property from billions of neural connections in the forebrain.

A pebble of awareness

Descartes also was wrong about consciousness. He defined the mind as nothing but conscious awareness. We now know that what occupies your conscious mind represents only a fraction of your brain's work at any given moment. Awareness sits atop unconscious mental processes like the visible hillock of an iceberg resting above a submerged and invisible mountain. While you read these words, your brain subconsciously and automatically is processing many things, including your body's position and balance (a "sixth sense" called proprioception); stimuli reaching your senses but not rising to the level of consciousness; and the nuances of language, such as the phrase "sixth sense" indicating something beyond the physical realm of the five you already know.

Scientists have yet to tease out the ways the brain acts outside consciousness to initiate a readiness potential for movement. Even exploring the subconscious presents a conceptual challenge.

"Consciousness is a much smaller part of our mental life than we are conscious of . . . How simple that is to say; how difficult to appreciate!" wrote psychologist Julian Jaynes. "It is like asking a flashlight in a dark room to search around for something that does not have any light shining upon it. The flashlight, since there is light in whatever direction it turns, would have to conclude that there is light everywhere." ■

The takeaway

Your mind is constantly at work in mysterious ways. When you think you have willed a particular part of your body to move, that thought comes after your brain already has ordered the movement. This time lag calls into question the nature of conscious will. If your brain initiates behavior that you can't discern until it's already under way, then your unconscious mind plays a powerful role in your feelings of being aware and making choices.

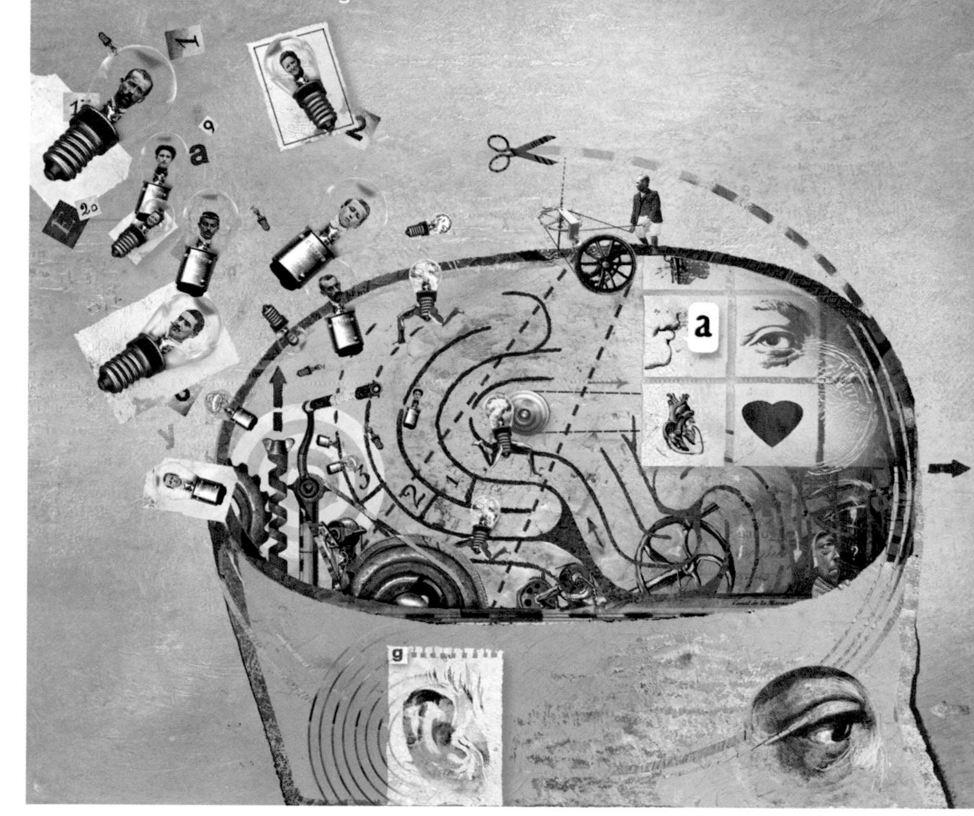

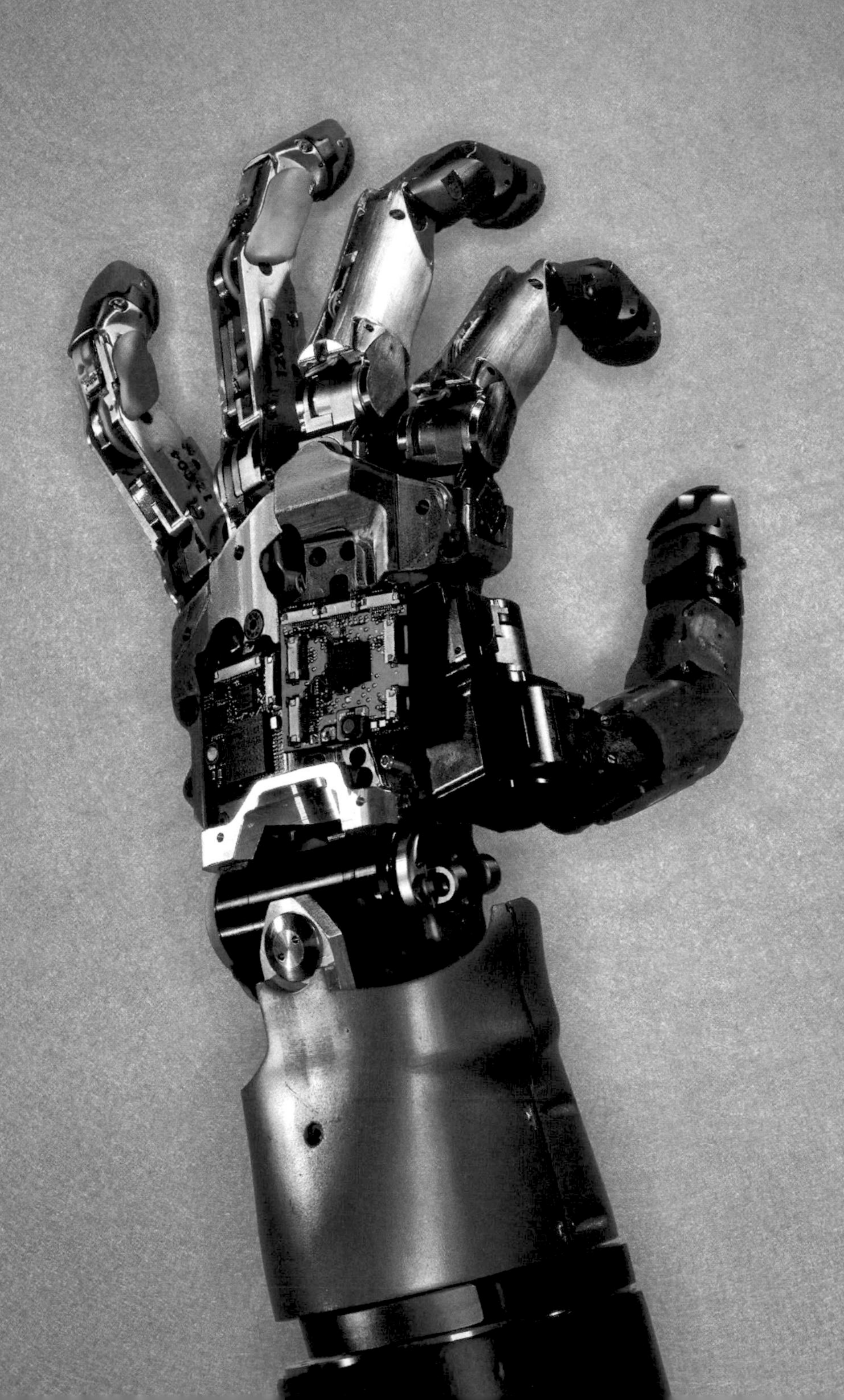

Artificial intelligence

Experiment

3.4

Choose

which of the following are easy for a computer to do:

Ride a bicycle.
Walk up a flight of stairs.
Play a grandmaster-level game of chess.
Divide a thousand-digit number by a seven-digit one.
Recognize a face.
Have a conversation about office politics.
Remember a list of 20 phone numbers.

Now turn the page.

What happened

Take another look at the list. Ask yourself: Which items could I do easily? If the task is easy for you, it's hard for a computer. And chances are, the reverse is also true.

The abilities humans take for granted—abilities learned as a child, say—give fits to computer designers. Recognizing a parent's face. Maintaining balance while walking up a flight of stairs. Having a rewarding conversation. Reading and understanding a book.

Accomplishing basic tasks such as these accounts for most neural circuitry. Small wonder. It's actually incredibly difficult to climb stairs. The brain must sense information by sight and touch, use it to command particular muscles to move, measure changes in balance, and continue to do so while a stream of neural feedback adjusts for the shifting environment from stair to stair. And it does so immediately and seamlessly.

On top of all that, you maintain an awareness of your body and your mind as you climb.

Compare this with artificial intelligence. "Despite all the computing power and effort that have gone into making computers intelligent, they still can't do what a three-year-old child can do," said Michael Gazzaniga, director of the SAGE Center for the Study of the Mind at the University of California, Santa Barbara. "They can't tell a cat from a dog."

Computers so far can only carry out narrowly defined programs, such as multiplying numbers or playing chess. Common human behaviors require using many skills, and many parts of the brain, at once. Brains are slower, but still much more sophisticated, than computers. ■

Step by step

Machines with silicon "brains" have great difficulty executing many tasks people find simple, such as climbing stairs.

Look closer

In 1956, scholars from Dartmouth, Harvard, IBM, and the Bell Telephone Laboratories called for a two-month retreat in New Hampshire to investigate a new concept: artificial intelligence.

"This study is to proceed on the basis of the conjecture that every aspect of learning or any other feature of intelligence can in principle be so precisely described that a machine can be made to simulate it," their proposal said.

The retreat's idea caught fire. In the 1960s, Herbert Simon, an attendee of the original conference, predicted, "Machines will be capable, within 20 years, of doing any work a man can do."

Obviously, machines haven't lived up to predictions. To be fair, there have been some computing advances that have freed people from certain tasks. Industrial robots do much of the repetitive work of mass production. Voice-recognition computers route calls at telephone centers. And computers crunch numbers very well.

Strong AI

But no machine has developed *strong AI,* a term fashioned by philosopher John Searle at the University of California, Berkeley. Strong AI wouldn't just mimic a particular skill associated with intelligence. It would have the cognitive states of a mind.

Robot chess

A boy plays chess against a robot at a 2007 auto exhibition in Moscow. A chess program fits a computer's abilities.

> By the 2030s, the nonbiological portion of our intelligence will predominate. Ray Kurzweil

Contrast this with weak AI—the use of software for solving particular tasks. Using weak AI, IBM's Deep Blue computer beat the world chess champion in 1997. It executed all the right moves but had no idea it had won.

Neural researcher Jeff Hawkins believes science will begin to approach strong AI when computers more closely model the human brain. The neocortex is not a collection of specialized hardware. Instead, it is highly integrated. Every part of the neocortex resembles every other part, leading to the conclusion that each part must be doing roughly the same job. Certain regions create different end results—vision arises in the visual cortex, for example—not because of variance in processing, but because of variance in

Made to serve

A Japanese robot pours tea from a bottle to a cup. The computing power of robots has reached the stage of small animal brains but has a long way to go to mimic a human.

input. Thanks to the brain's intricate neural networks, information arising in one region easily moves to other regions in a variety of ways, making the whole greater than the sum of its parts.

The brain's trillions of interneural connections, plus its ability to make new connections, fundamentally separate it from a machine. In addition, millions of years of evolution have selected for certain un-machinelike abilities in the brain as important for survival. Parents and children recognize each other's faces for their own good. Newborn brains learn and process language because that skill helped early humans gather food and fight threats. Meanwhile, humanity's ancestors didn't need much mathematical calculation for basic survival. That's a key reason why that particular skill remains relatively weak in the human brain, compared with the power of computers running mathematical programs.

Becoming conscious

Consciousness appears to be a relatively recent evolutionary development; animals don't need

CASE STUDY

Language machine

Engineers trying to build machines that mimic human behavior have struggled with language. Words can be subtle, and understanding them becomes complex as they get layered with nuances of spoken emphasis, body language, and syntactical constructions that give hints to meaning. For example, a slight rise in tone at the end of a statement turns it into a question.

In 2010, IBM announced it had made a breakthrough with Watson, a computer that rivaled humans' language-recognition ability in written form.

The company fed Watson's memory vast amounts of data from books, dictionaries, encyclopedias, and other texts. When asked a question, Watson mined the data using algorithms that made statistical models of the most likely answers, based on words' definitions and common associations. Watson then chose and articulated a single answer.

IBM entered Watson in the television answer-and-question program *Jeopardy!* in February 2011. Watson received the text of clues electronically at the same time the human contestants heard and read them. Watson rang in only if its algorithms reached a high degree of confidence in the answer. Using the game's required question form for its answers, it then "spoke" with a voice synthesis program.

Watson proved formidable in warm-up scrimmages against test subjects. "He plays to win," graduate student Samantha Boardman told the *New York Times*. Most contestants referred to Watson as "he."

Watson stumbled a bit with a test category about IBM. But it gave the correct response when told, "It's the last name of father and son Thomas Sr. and Jr., who led I.B.M. for more than fifty years." Quoth the machine, "Who is Watson?"

I visualize a time when we will be to robots what dogs are to

it to survive. Scientists now consider consciousness an emergent property, arising out of the human brain as it evolved a huge frontal lobe and its total processing power passed some unknown boundary.

What might that boundary be for a machine?

Considering the complexity of memory and circuitry today, computers appear to have the intellectual power of insects. Robotics researcher Hans Moravec noted the similarities in *Scientific American*. Ants easily follow scent trails but struggle to find their way if the trail gets disrupted. Likewise, industrial robots sense guide wires buried in concrete floors or scan bar codes on walls to navigate inside a room, but they can't navigate across open ground.

Many industrial robots, as well as modern desktop computers, have the processing abilities of a small fish. That's about 10,000 or more million instructions per second (MIPS), a standard measure of computing power. Machines with a lizardlike 20,000 MIPS or more also are already here. Robots with such a processing level can do a variety of simple chores but are limited by their application programs. "Unable to adapt to changing circumstances, they will often perform inefficiently or not at all," Moravec wrote.

Humanlike robots

A machine that matches the human processing power of 100 million MPS would have the abilities to think abstractly, to handle contradictions, and to generalize. It would move freely and match a human's

"I'm sorry, Dave"

The fictional HAL 9000 computer speaks to astronaut Dave Bowman (actor Keir Dullea) in the 1968 movie *2001: A Space Odyssey.* HAL, short for Heuristically programmed ALgorithimic computer, acts like a human, for good and ill.

Computer History Museum home page: **http://www.computerhistory.org/**
Past, present, and future of computers.

humans, and I'm rooting for the machines. Claude Shannon

intellectual power. Moravec anticipates crossing that boundary by 2040.

How will we know when a machine can think? It's a difficult question, as nobody can directly examine the thoughts of another. Alan Turing, founder of computer science, proposed his now-famous Turing test in 1950: Engage in a conversation with two unseen partners, one human and one a machine, and if you can't tell which is which, the machine passes the test. Some have criticized the experiment as measuring behavior instead of intelligence, but no matter. No machine has yet aced a Turing test.

The best chat bots, such as ALICE and ELIZA, are getting closer, but nobody would claim that any such computer programs designed to simulate conversations are intelligent. Neither is Watson, the IBM supercomputer that handily defeated *Jeopardy!* champions Ken Jennings and Brad Rutter in a 2011 showdown. Watson did not have the ability to know it had won. ■

The takeaway

Compared with the human brain, computers are fast but dumb. Thanks to its massive, complex integration of neurons, the brain can learn and adjust to handle a huge variety of tasks requiring the processing of multiple streams of information. The brain makes child's play of climbing stairs or understanding a book, tasks that baffle machines. But humanlike computers may be only a few decades away. A big breakthrough will occur when computers reach the point of modifying and even improving their own designs. Then they might find their own methods of mastering simple human tasks.

Being in pain

Experiment **3.5**

Imagine

you're a secret agent being tortured. To play mind games, your captor lets you choose: You can suffer ten seconds of pain that rates a 9 on a scale of 1 to 10, or you can opt for ten seconds of number 9 pain, followed by ten more seconds of pain that hits a 7 on the same scale.

Which do you choose?

Why did you make that choice?

Now turn the page.

What happened

Choose the second option. You'll feel better about it afterward. This is not just an imaginary scenario.

Psychologist Daniel Kahneman and colleagues conducted a very similar experiment—but without the sinister elements of a spy movie. Volunteers subjected themselves to two ordeals. In one, they thrust a hand in water that registered 14°C (57.2°F) for 60 seconds. That's cold enough to cause discomfort but not severe pain. In the other trial, they thrust a hand into 14°C water for 60 seconds and then kept it there for an extra 60 seconds while the water slowly warmed to 15°C. Some experienced the short

More or less

Enduring just a few more seconds of ice-cold torture—as long as it doesn't hurt *quite* so badly—will create a slightly more pleasant memory of the experience than ending when things are at their worst.

Daniel Kahneman home page: **http://www.princeton.edu/~kahneman/**
Publications and lectures on decision-making by the Nobel Prize winner.

ordeal first; others started with the longer test.

Later, Kahneman and his team asked the volunteers which ordeal was more painful. Counterintuitively, most chose the shorter one. They also said that if they had to choose a third ordeal, they would ask for the longer one.

Kahneman describes this phenomenon as the peak-end rule. The brain evaluates painful and pleasurable experiences by the moment of greatest intensity of sensation plus the quality of the final sensation. Duration of pain or pleasure means almost nothing. People remembered the two-minute ordeal as less painful than the one-minute version because it ended slightly better.

The peak-end rule has been confirmed in a variety of experiments and has real-world consequences. As you plan a vacation, keep in mind that you'll remember it more fondly if it contains a peak experience and ends with great pleasure. That means a two-week vacation could very well produce fonder memories than a three-week one. ■

Look closer

Two hundred years ago, British philosopher Jeremy Bentham observed that people execute a kind of pleasure-pain calculus to decide how to act.

They sum the expected elements of pleasure and pain, and then they seek to maximize the former and minimize the latter. Bentham called this philosophy utilitarianism. Memories of pain and pleasure, which guide future behavior, become "remembered utility."

Kahneman's remembered-utility experiment runs counter to Bentham's ideas. If people weighed pain like bricks, surely the memory of a two-minute plunge into ice water that had a slightly less painful ending would outweigh the one-minute alternative.

The complexity of remembering pain explains why this is not true.

Sensations of pain arise in specialized nerve cells called nociceptors. They sense harm to the body's tissues through such stimuli as heat and cold, pressure, cutting, and crushing. In response, they send the brain electrical impulses that get interpreted as pain. Due to natural selection, painful sensations are encoded as memories, which better enable individuals to survive and to pass on their genes. If you accidentally burn your hand in a fire, memories of the experience likely will prevent you from a second dangerous exposure.

A memory is not as the same as the original experience, however. And thank goodness. Imagine refeeling the pain of a bad burn each time you remember it.

Nor is memory purely physical. Sensations combine with emotional and cognitive stimuli during encoding. The brain apparently draws from peak and end to summarize the entire experience in memory.

Applications and ethics

The results of Kahneman's research could have applications in medicine. Patients' willingness to undergo physical exams depends heavily on how they evaluate previous ones.

CASE STUDY

Feeling no pain

When he was a boy in the late 12th century, Baldwin IV, heir to the throne of the Christian kingdom of Jerusalem, played a game with his friends. They pinched and scratched one another to determine who was toughest.

"The others evinced their pain with yells, but, although his playmates did not spare him, Baldwin bore the pain altogether too patiently, as if he did not feel it," recalled Baldwin's tutor, historian William of Tyre. "When this had happened several times, it was reported to me. At first I thought that this happened because of his endurance, not because of insensitivity. Then I called him and began to ask what was happening. At last I discovered that about half of his right hand and arm were numb, so that he did not feel pinches or even bites there."

Baldwin had leprosy, a rare, infectious disease that killed his pain. It ate away at his nerves but did not prevent him from ascending the throne. For the rest of his short life, he ruled wisely and well as the "leper king" who brought relief from war with Muslims.

A variety of genetic diseases also causes hereditary sensory and autonomic neuropathies, which inhibit sensation. Some block pain.

Six related children in Pakistan described in the journal *Nature* (2006) inherited a pain-free life. All had damaged lips from chewing. Most had suffered bone breaks diagnosed only after they began limping. Analysis showed that their DNA prevented a particular protein from performing a key function in signaling pain.

Peak and end

Your overall memories of an event rely upon two key moments: that of greatest intensity, such as the powerful emotion at right, and that of the end.

In one experiment, Kahneman, Joel Katz, and Donald A. Redelmeier asked men who were having medically required colonoscopies how they felt moment by moment during the procedure. Afterward, they asked them to rate the overall experience.

It would be difficult to describe the exam as pleasant. A long, flexible tube containing a tiny camera is inserted through the rectum and pushed along the inside of the colon. One group of patients in the experiment had a standard colonoscopy. A second group had the same procedure—up to a point. After finishing, the doctors left the tube in place for 20 seconds. The second group reported that the sensation remained unpleasant, but it wasn't as bad because the tube didn't move. The doctors then removed the tube.

Both groups had the same peak experience of pain, but the second group had a milder concluding experience. Their self-reported memories indicated less pain overall. During the next five years the second group proved more likely to visit the doctor for follow-up colonoscopies.

This raises significant ethical questions. Should a doctor continue to cause pain in order to modify memories of a medical procedure? And does that conflict with the Hippocratic oath to do no harm? ■

*The takeaway

As you remember how something felt on the spectrum of pain and pleasure, your brain places high importance on the moment of greatest intensity and the moment at the end of the experience. This sample then stands as the memory of pain or pleasure for the entire experience. Ending on a good note plays a key role in encoding memories that heighten the feelings of good times and lessen the feelings of bad ones. So, when planning your next vacation, it would make sense to schedule an amazing ending.

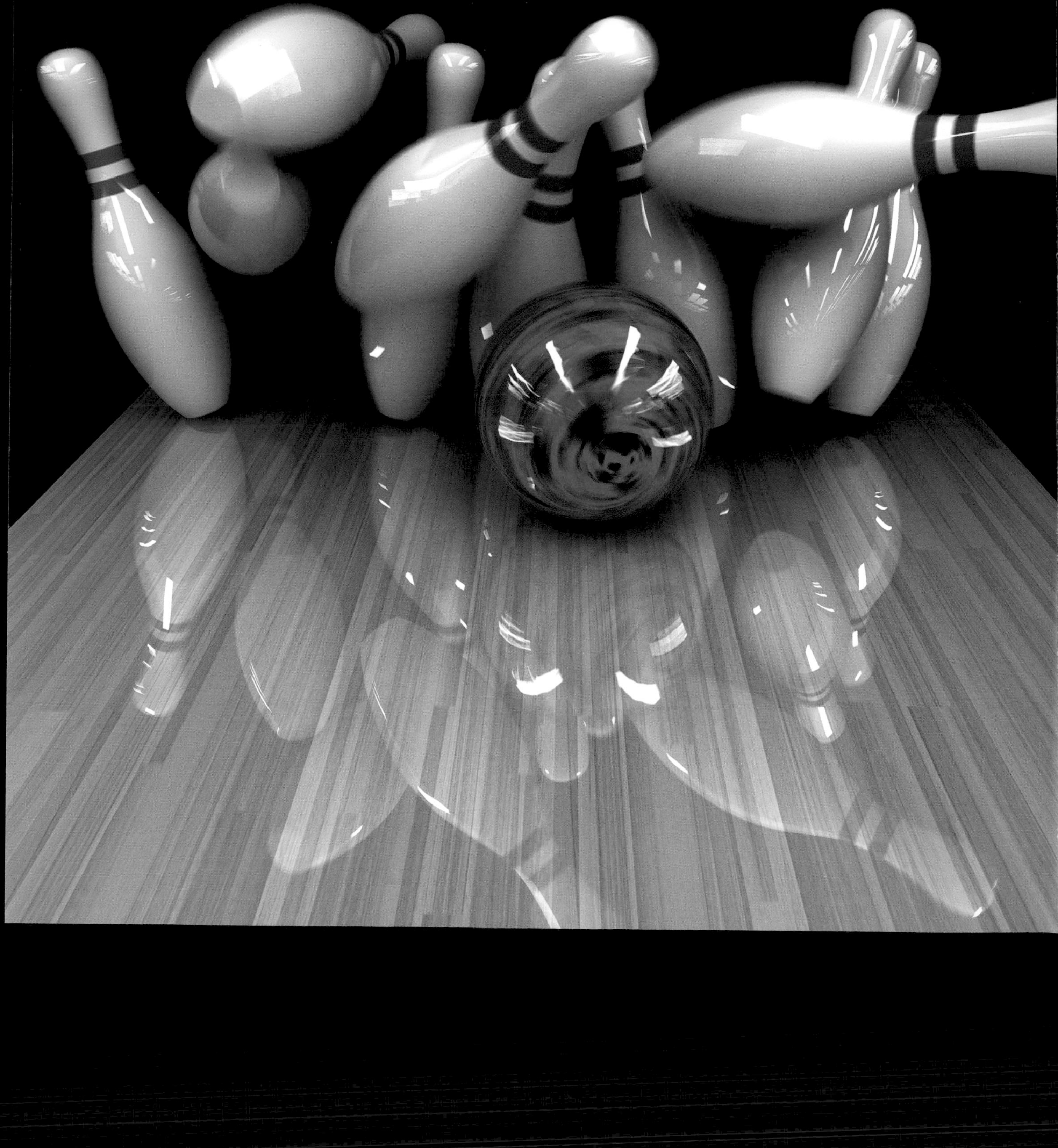

Being **happy**

Experiment **3.6**

Predict

the length and intensity of your feelings if the following happened today:

You won the lottery.

Your salary doubled.

Your favorite sports team won a big game against its chief rival.

Someone close to you, such as a spouse or child, died.

You got an unexpected refund of $20, or you lost $20 from your wallet.

You got fired.

Now turn the page.

What happened

In the case of winning the lottery or doubling your salary, did you imagine bliss stretching for months or years?

Or did you foresee ongoing devastation from being fired? Either way, you'd likely turn out to be wrong.

When you predict happiness or sadness, you employ affective forecasting. And you're probably bad at it. Humans have what psychologists call an impact bias toward overstating the intensity and duration of emotional states in response to events they cast as good or bad.

"Nothing in life is quite as important as you think it is while you are thinking about it," psychologist Daniel Kahneman said. The brain's emotional circuitry is well grounded; it turns crises into blips. Events viewed as life changing end up being more like the loss or gain of $20. Even the pain of losing a child or spouse usually fades with the intensity of new experiences.

Foggy forecasts exist for several reasons.

First, the brain falls victim to what Kahneman calls the focusing illusion. Happiness arises from many things, but people tend to exaggerate the importance of a single factor, such as money or love.

Second, people overestimate how they will feel in the future because they underestimate their own resilience. They do not realize how quickly they return to an emotional set point—an equilibrium that fluctuates and then stabilizes.

Daniel Gilbert provides a psychological explanation for misreading the joy of achievement: The pursuit or anticipation of a goal feels as good as, if not better than, when the event actually occurs. He likens this to "getting double the juice from half the fruit." ■

Mood forecast

Into each life a little rain must fall . . . and a healthy brain is wired not to make too big a deal about life's highs and lows.

! Look closer

Much scientific and anecdotal evidence supports the existence of the focusing illusion and the preponderance of inaccurate affective forecasting.

The focusing illusion stands out when you pay close attention to things that seem particularly positive or negative. In their book *Happiness: Unlocking the Mysteries of Psychological Wealth*, Ed Diener and Robert Biswas-Diener use living conditions in Indian slums as an example. Typical Western visitors to the slums of the big coastal cities likely feel overwhelmed by the widespread poverty, hunger, unemployment, and lack of health care. They conclude that slum life must be miserable. However, slum residents rate their lives as neutral or even slightly positive on a scale balancing unhappiness and happiness. "If this sounds surprising to you," the authors wrote, "it is because you, like us, overlooked the close family relationships, soccer games, religious celebrations, new babies, weddings, card games, and many other positive elements that are a daily part of slum life."

In the United States, Kahneman and colleagues found the focusing illusion when they surveyed Californians and Midwesterners. When both groups responded to surveys, they predicted Californians would be the happier of the two. Respondents favored the salience of pleasant weather. But in fact, when rating their own happiness, Californians and Midwesterners scored the same. The Midwest offers much that doesn't get counted in a weather forecast, and California has its troubles.

In everyday life, the focusing illusion influences consumer decisions that involve comparisons. If you go looking for a new computer because you want one with a bigger screen and more memory, you'll no doubt find electronics store display stands filled with huge computer monitors side by side, as well as advertisements for mammoth operating systems. You probably

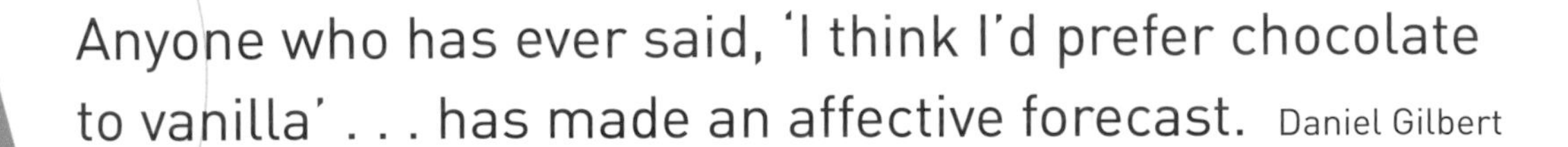

> Anyone who has ever said, 'I think I'd prefer chocolate to vanilla' . . . has made an affective forecast. Daniel Gilbert

will tend to focus on those two qualities as you decide which computer to buy, perhaps ending up with more than you really need or want. Evaluated by itself in your office, without comparison brands beside it, a smaller computer might have served you just as well, but the focusing illusion led you toward more expensive purchases.

Quick changes

Research indicates that strong feelings tend to fade quickly; humans are more resilient than they know. In one study on notoriously moody teenagers, it took only an average of 45 minutes for them to return to their emotional baseline after experiencing an intense high or low feeling.

Psychologists Daniel Gilbert and Timothy D. Wilson decided to study the impact of ostensibly life-altering decisions on adults and found the emotions weren't as powerful as expected. They asked young professors seeking tenure how they would feel if they did or didn't get it. Typically, beginning professors must achieve tenure in a few years on the basis of their teaching, research, and service, or else lose their contracts and seek work elsewhere. The professors said they would feel long-term elation or devastation. But that's not what happened later. Both those who did and those who did not get tenure felt a brief emotional surge and then got on with their lives.

Self-reports of lottery winners provide some of the starkest evidence of the transitory nature of bliss and the weakness of affective forecasting. Winning a huge cash prize naturally provokes a giant jolt of happiness, but it doesn't last. Money turns out to be insufficient for long-term joy. A study of the million-dollar winners of the New Jersey Lottery reported that nearly all of them suffered harassment and threats, and many lived in fear. Friends and family members came to resent the winners' good luck, and strangers bombarded them with requests for handouts. "They have won the battle against poverty and deprivation, but are losing the war," wrote an interviewer of lottery winners. The winners "are financial successes but social and psychological casualties." The first winner of the $1 million Maryland State Lottery in 1973 was living in a two-bedroom apartment and did not own a car when a reporter tracked him down in 1993. He was adamant that he would have been better off if he had not won.

Emotional ups and downs

Humans (and other animals) appear to be genetically programmed not to get emotionally stuck. Wilson likens happiness to blood pressure. It fluctuates for the good of the body, but it can't remain high or low for very long without doing harm.

Money = happiness?

Studies of lottery winners have shown that many fail at affective forecasting—they don't experience the long-term bliss they expected from sudden wealth. After an initial surge of joy, winners return to an emotional set point, or they drop below it because of financial complications.

PRIZE WINNER
BILLION DOLLARS and 00/1
1016

> You can live as if nothing is a miracle; you can live as if everything is a miracle. Albert Einstein

BREAKOUT

Test your happiness

Below are five statements that you may agree or disagree with. Using the 1 to 7 scale below, indicate your degree of agreement with each item by checking the circle below the appropriate number. Please be open and honest in your responses.

7. Strongly agree
6. Agree
5. Slightly agree
4. Neither agree nor disagree
3. Slightly disagree
2. Disagree
1. Strongly disagree

In most ways my life is ideal.

1	2	3	4	5	6	7
❍	❍	❍	❍	❍	❍	❍

The conditions of my life are excellent.

1	2	3	4	5	6	7
❍	❍	❍	❍	❍	❍	❍

I am satisfied with my life.

1	2	3	4	5	6	7
❍	❍	❍	❍	❍	❍	❍

So far I have gotten the important things I want in life.

1	2	3	4	5	6	7
❍	❍	❍	❍	❍	❍	❍

If I could live my life over, I would change almost nothing.

1	2	3	4	5	6	7
❍	❍	❍	❍	❍	❍	❍

Scoring: 31–35, extremely satisfied; 26–30, satisfied; 21–25, slightly satisfied; 20, neutral; 15–19, slightly dissatisfied; 10–14, dissatisfied; 5–9, extremely dissatisfied.

Source: "Satisfaction with Life Scale" by Ed Diener, Robert A. Emmons, Randy J. Larsen, and Sharon Griffin, as noted in the 1985 article in the *Journal of Personality Assessment.*

The body benefits from reacting emotionally; changes in pulse, adrenaline, and blood pressure, for example, are crucial reactions to fear that prepare the body for survival.

However, imagine getting stuck in the pit of fear, or in the sweaty, giddy pleasure of ecstasy. It not only would be exhausting, but also would likely prevent new emotions from registering. "If people are still in a state of bliss over yesterday's success, today's dangers and hazards might be more difficult to recognize," Wilson wrote. "In short, it is not good for us to be depressed or euphoric for long."

There apparently are limits to the level of happiness humans can experience, as well as its duration. Wilson sees good and bad in this. The good is that neural programming prevents you from staying too long in a particular emotional state, although the incidence of chronic depression is proof that this internal mood-stabilizing system can break down. The bad is that you might want to have a blissful mental state, such as the blush of first love, continue much longer than it does. Unfortunately, Wilson said, people have "physiological and psychological mechanisms that, basically, rain on their parades."

Temperament no doubt plays a role in the setting and resetting of mood. "Happiness and misery depend as much on temperament as on fortune," said 17th-century French writer François de La Rochefoucauld. This quality of a personality baseline for temperament suggests that a predisposition for happiness or sadness is at least partly hereditary, and indeed researchers have found some genetic evidence of a predisposition toward happiness.

Professor Yoram Barak of Tel Aviv University's Sackler School of Medicine has been seeking a "happiness gene" in his studies of twins. If DNA is partly responsible for a person's set point for happiness or sadness, there should be a high correlation in the set points of twins, whether reared together or apart. Barak said in 2009 that based on data so far, he believes about 50 percent of happiness

CASE STUDY

Phineas Gage

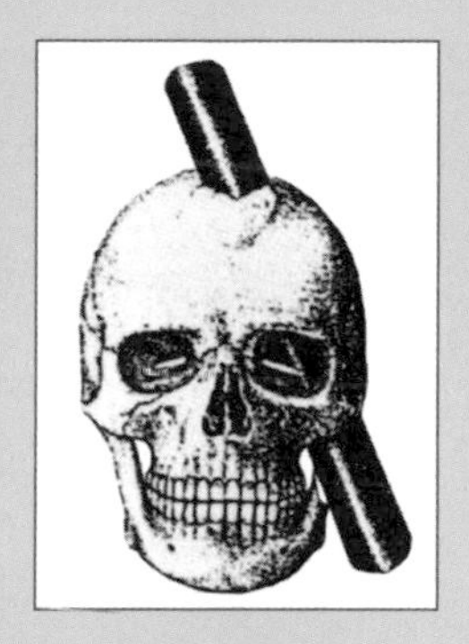

Predictions of future happiness arise in the forebrain. They play a crucial role in everyday decisions, according to Harvard University psychology professor Daniel Gilbert. When you decide between oatmeal and yogurt for breakfast, or between getting up early to exercise or staying in bed an extra hour, you do so because you predict that one will bring you greater psychic rewards in the immediate or more distant future.

But how might your life change if you couldn't decide how various choices might make you feel?

One of the most famous patients in the history of neuroscience provides a clue.

Phineas Gage was working along a Vermont railroad track in 1848 when a freak explosion shot a 13-pound metal bar through his left cheek and out the top of his skull. The accident destroyed much of Gage's prefrontal cortex and the vision in his left eye. But Gage astonished everyone by initially showing few ill effects. He remained lucid and able to walk. His skull torn open and a portion of his brain blown out, Gage calmly told a doctor, "Here is business enough for you."

Gage's physical wounds closed, but mental ones soon opened. He shifted from being a capable and efficient crew foreman to an impatient, profane, drunk, and unsociable fellow. He wasn't happy.

More than his personality had changed, however. Damage to his frontal lobe had left Gage incapable of sound affective forecasting. A doctor who examined Gage observed him "devising many plans of future operation, which are no sooner arranged than they are abandoned in turn for others appearing more feasible."

Gage toured for a while with P. T. Barnum's circus, but he continued to find little, if any, happiness. He died 12 years after his accident, apparently of complications from seizures.

When something good happens, people are happier for a longer

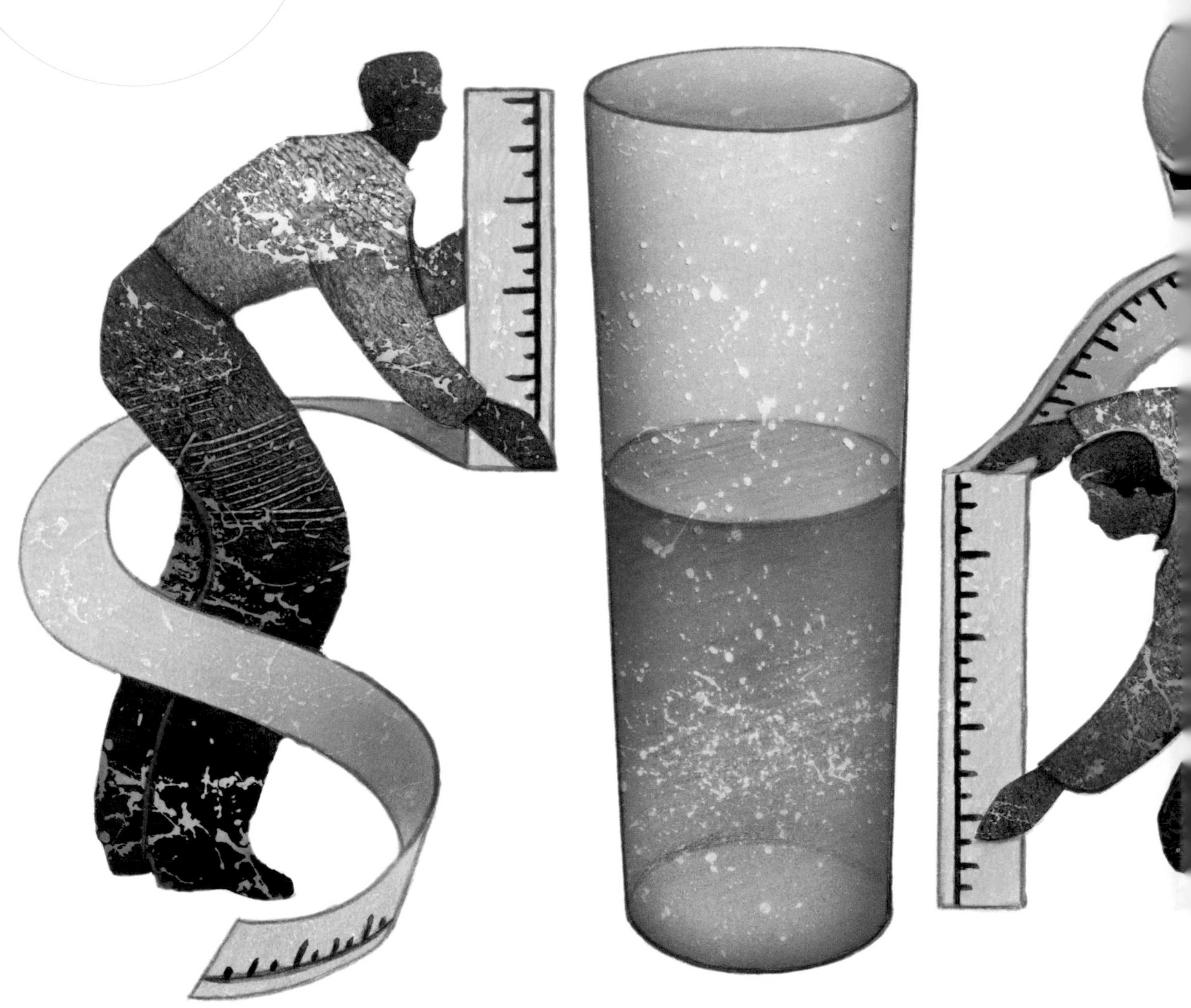

time when they don't fully understand the reasons why it happened. Daniel Gilbert

is genetic. And the other 50 percent? Barak's studies suggest that the power of positive thinking, promoted in positive psychology workshops, can raise happiness in the long term.

Reframing events can have a significant impact. A painful divorce, once completed, can be reframed as a liberating event and the start of a new life. Losing television reception in a snowstorm can be an excuse to enjoy family games, conversation, or a good book. Having your car in the repair shop can give you a reason to enjoy walking or biking to work, or to check out the environmentally friendlier option of public transportation. ■

Which half?

Some people are born to see a glass as half full. Heredity accounts for about half our capacity for happiness.

*The takeaway

For the good of your physical and mental health, your emotional states fluctuate but generally return to a baseline, much the same way a thermostat controls heating and cooling to keep a room at a comfortable temperature. Strangely, people tend to be unaware of their own emotional resiliency. They overestimate the intensity and duration of future emotional states, and they place too much emphasis on how one particular experience may affect their mood.

A FINAL THOUGHT

Perhaps this book has made you a bit more cautious, a bit more uncertain, about how you view the world and your interactions within it. If so, you're not alone. The brain's complexity still staggers the imaginations of those who have devoted their lives to trying to understand it better. There is no subject so maddeningly complex, yet so vital to understanding the human experience, as the functioning of the brain. • The main roadblock to understanding the brain is that it doesn't lend itself to direct study in the same way as an arm or kidney. Western science traditionally favors the process of reduction: It takes apart a tricky problem into the smallest of pieces, figures out how they work, and then reassembles the bits to create the whole. To understand what makes an engine convert the chemical energy of gasoline into the kinetic energy of your car moving down the highway, you could (if you had lots of time and a high tolerance for tedium) strip the car into individual parts, analyze each one until you understand it, and put it back together again.

Reality is merely an illusion, albeit a very persistent one. Albert Einstein

Endless climb

Page 216: A three-dimensional model, inspired by Dutch artist M. C. Escher's famous staircase, confuses the issue of which way is up.

The brain, on the other hand, is more than the sum of its parts. In fact, it doesn't even have parts in the same manner as your car. Some areas—such as the visual cortex for processing light and the amygdala for registering fear—appear to possess relatively specialized functions despite the fact that one neural network looks pretty much like any other. Furthermore, each network, working on its own to process such data as sensation, emotion, and thought, does not function in a vacuum. Each adds its drop of water to the spreading pool, until consciousness bursts the bank and the brain becomes aware, and knows it is aware.

The poor reductionist finds that it is only when the myriad regions of the brain work together, storing, transmitting, and modifying bits of information, that the really interesting and amazing phenomena of the brain emerge.

Consciousness. Personality. Memory. Reason. Emotion. And all of the other seemingly magic tricks that begin as simple electrical pulses jumping from neuron to neuron.

So what's a neuroscientist to do?

As you have seen in this book, a common method of analyzing the brain is to observe what happens when things go wrong, or when the brain produces information that reason cannot refute. Insights into optical illusions such as the Müller-Lyer, or arrow, illusion and the Hermann grid reveal ways in which the eye processes line and light. Cognitive puzzles such as those involving issues of choice and conscious will reveal the brain's preference for order and control. Damage to specific brain regions that impairs or destroys particular functions, such as the ability to read or see color, leads neuroscientists farther along the path to understanding how those functions arise in the first place.

In all of the above cases, scientists argue that the brain's quirks exist for a reason. As the biologist Theodosius Dobzhansky observed, "Nothing in biology makes sense except in the light of evolution." The human brain, pinnacle of billions of years of evolution, is a biological organ. Thus, there would appear to be an evolutionary advantage, an enhanced chance of survival, inherent in everything from the retina's

lateral inhibition, which explains the Hermann grid, to the unequal weighing of losses and gains.

As this book has shown, your brain does play tricks on you. You should be mindful, and thankful, for that means your brain isn't a mere computer. The human brain has evolved and matured into the most complicated and profound object in the universe, processing information in ways that computers may never duplicate. Yes, the brain is right most of the time, or else humans would not still walk the Earth. But it's revealing, humbling, and perhaps even fun to know that, like humanity itself, the brain isn't perfect. ■

In or out

In a trick of the brain interpreting shadow, footprints in sand may appear inside out. Rotate book 180 degrees and see what happens.

Glossary

Agnosia. Inability to recognize objects. Visual agnosia is the inability to recognize things by their shapes and colors despite having functional eyes and memory.

Agraphia. Loss of the ability to write due to damage to the brain's language areas. Typically caused by stroke.

Alexia. Loss of the ability to read; caused by brain damage.

Amacrine cells. Type of retinal cell that transmits signals laterally. They share synapses with ganglion and bipolar cells.

Amygdala. Region at the back of the brain, below the cortex, involved in emotional response and olfactory processing.

Attention. Seeking and focusing on sensations that are of interest; causes them to be more strongly processed than other stimuli.

Autonomic nervous system. Portion of the peripheral nervous system that automatically controls respiration, heartbeat, digestion, and other functions that operate nearly entirely below the level of consciousness.

Axon. Neuron extension that conducts electrical impulses. Also called the nerve fiber.

Bipolar cells. Cells that receive and process electrical impulses from rods and cones and send them to ganglion cells.

Blind spot. Small area where the optic nerve leaves the back of the eye. It contains no photoreceptors, so tiny objects are invisible if light reflected from their surfaces strikes this region.

Brain stem. Region at the back of the brain adjacent to and structurally contiguous to the top of the spinal cord. Includes the medulla oblongata, pons, and midbrain. Functions located in the brain stem include regulation of breathing, digestion, heartbeat, blood pressure, sleep-wake cycles, and alertness.

Cerebellum. Brain region above the brain stem; responsible for coordination of voluntary muscle movement and balance. Stores memories of learned, complex motions and allows unconscious processing of actions that previously required concentration to perform.

Cerebral cortex. Two-millimeter-thick covering of neurons atop the cerebrum. Associated with many functions of cognition, such as memory and thinking.

Cerebrum. Largest and evolutionarily newest portion of the human brain. Consists of the frontal, temporal, parietal, and occipital lobes, all of which exist in two hemispheres, left and right. Home to perception, judgment, decision-making, imagination, language, and other complex functions.

Cognitive dissonance. Uncomfortable psychological condition caused by holding two or more conflicting thoughts or feelings at the same time. Typically resolved by changing behavior or ideas to bring them into harmony.

Cones. Cone-shaped photoreceptors in the retina; responsible for vision at high levels of illumination, in color and in great detail.

Cornea. Transparent focusing element at the front of the eyeball. It is the first structure through which light passes on its way to the retina.

Corpus callosum. Thick band of neural fibers that connects the brain's left and right hemispheres and allows them to communicate information.

Declarative memory. One of two types (the other is procedural) of long-term memory, in which a person recalls facts or events.

Dendrite. Branching projection of a neuron. Receives electrical impulses from another neuron and sends them to the cell body. Most dendrites share synapses with nearby axons.

Diencephalon. Portion of the midbrain that includes the thalamus and the hypothalamus, which play key roles in the regulation of the autonomic nervous system.

Dyslexia. Broad term covering a variety of conditions that impair a person's ability to read. Associated with the mental processing of words, it is considered a learning disability, not an intellectual one.

Episodic memory. One of two types (the other is semantic) of declarative memory; consists of memories of autobiographical events.

Excitatory response. Increased rate of firing in a nerve fiber in response to stimuli.

Fovea. Tiny region of the retina containing only cone cells, and thus well-detailed color vision. Automatically stimulated by focusing the eyes on an object.

Frontal lobes. Largest lobes of the human brain. They lie at the front of the brain and are associated with executive function.

Ganglion cells. Neurons in the retina that receive electrical signals from bipolar and amacrine cells, and extend their axons out the back of the eyeball as the optic nerve.

Hippocampus. Brain region inside the medial temporal lobe. Takes its name from its resemblance to a seahorse. Crucial for formation of long-term memories.

Horizontal cells. Retinal cells that transmit electrical impulses from side to side and share synapses with rods, cones, and bipolar cells.

Lateral geniculate nucleus. Region of the thalamus that receives visual information from the optic nerve, processes it, and relays it to the visual cortex.

Lateral inhibition. Neural inhibition transmitted from side to side, as in the retina. Causes perception of greater contrast in adjacent objects; crucial for recognition of edges.

Lens. Transparent element of the eye that changes shape, to shift focus, in response to muscular contractions.

Mesial prefrontal cortex. Region of the forebrain that processes reward outcomes, including rewarding stimuli such as attractive faces, monetary gain, and pleasant music.

Motor cortex. Portions of the cerebral cortex, stretching roughly from the top of the head to the ear on both sides, that plan, execute, and control voluntary movement.

Neural circuit. A number of neurons connected by synapses and sharing a function.

Neuron. Basic impulse-conducting unit of the nervous system. The complexity of the brain arises from the trillions of connections among tens of billions of neurons.

Neuron theory. Theory proposed in the 1800s, now universally accepted, that the nervous system consists of individual cells that do not physically touch one another but are rather functionally linked through synapses.

Neurotransmitter. One of many kinds of chemicals stored in the membrane of a neuron. It is released in response to an electrical impulse, travels across the synapse, and stimulates the neuron on the other side.

Nociceptors. Neural receptors that fire in response to pain.

Nucleus accumbens. Brain region behind the prefrontal cortex; associated with rewards, pleasure, and addiction.

Occipital lobes. Smallest lobes of the brain, at the rear of the head. Home to processing of visual stimuli and dreams.

Optic nerve. Bundle of nerve fibers (axons) that carries impulses from the retina to the lateral geniculate nucleus and other brain structures.

Orbitofrontal cortex. Region of the prefrontal cortex involved in assessing rewards and in decision-making. In the human brain, it consists of Brodmann areas 10, 11, and 47.

Pareidolia. Tendency of the brain to see familiar patterns in unfamiliar objects. It is particularly strong for perception of faces.

Parietal lobe. Portion of the brain above the occipital lobe. Functions include synthesizing information from a variety of sensory systems.

Perception. Conscious sensory stimulation.

Prefrontal cortex. Forward-most portion of the frontal lobes of the human brain, anterior to the motor and premotor cortices. Responsible for complex behavior unique to humans, including long-term planning, proper social behavior, and matching thoughts and actions to specific goals.

Procedural (motor) memory. One of two types (the other is declarative) of memory, in which a person unconsciously recalls learned motor skills, such as riding a bike or typing.

Prosopagnosia. The inability to recognize faces, also called face blindness.

Retina. Network of cells lining the inside back surface of the eyeball. It includes light-sensitive cells that generate electrical signals as well as cells that process those signals.

Rods. Rod-shaped photoreceptors in the retina. They function at low light levels but cannot generate sharp images.

Saccade. Small, quick eye movement, as when reading.

Salience. Relative prominence of information stored in memory, as measured by the ease with which it can be recalled.

Semantic memory. One of two kinds (the other is episodic) of declarative memory, consisting of memories of facts and meanings independent of personal experience.

Simultaneous contrast. Phenomenon of perception in which surrounding a color with another color causes an apparent change in the surrounded color.

Striate cortex. Also called visual cortex, the portion of the occipital lobe that receives and processes visual information. So named because it appears to have stria, or layers.

Subliminal priming. Priming that occurs below the level of conscious awareness.

Synapse. Fluid-filled gap between two neurons, across which neurotransmitters carry information electrochemically.

Temporal lobes. Lobes on the sides of the brain, below the frontal and parietal lobes; associated with high-level auditory and visual processing, as well as memory.

Thalamus. Brain region between the midbrain and cerebral cortex. Relates sensations to the cortex and aids in creating states of consciousness and alertness.

V1–V5 regions. Portions of the visual cortex that process information relayed from the lateral geniculate nucleus and thus add layers of complexity.

Working memory. Ability of the brain to hold and manipulate information for complex tasks, such as reasoning and learning.

Further reading

Chabris, Christopher, and Daniel Simons, *The Invisible Gorilla and Other Ways Our Intuitions Deceive Us* (New York: Crown, 2010). An exploration of psychological misconceptions. Humorously uses case studies to undermine confidence in the certainty of sensation and memory.

Diener, Ed, and Robert Biswas-Diener, *Happiness: Unlocking the Mysteries of Psychological Wealth* (Malden, MA: Blackwell, 2008). A broad range of research on the causes of happiness and sadness, including practical advice on how to be happy as well as common misconceptions.

Gazzaniga, Michael S., *Human: The Science Behind What Makes Your Brain Unique* (New York: Harper Perennial, 2008). A tour of the human brain, including an examination of its functions' development through evolution, written for an ordinary reader. Examines the interplay of brain regions, consciousness, aesthetics, and social interactions.

Gilbert, Daniel, *Stumbling on Happiness* (New York: Vintage Books, 2006). A Harvard psychologist explains common mental heuristics, including errors of forecasting, that interfere with achieving a happy, well-balanced life.

Gregory, Richard L., *Eye and Brain: The Psychology of Seeing* (NJ: Princeton University Press, 1979). A short, somewhat technical description of the basics of visual perception.

——, *Seeing Through Illusions* (England: Oxford University Press, 2009). A discussion of how the brain interprets what the eyes see. Organized by type of illusion.

Hoffman, Donald D., *Visual Intelligence: How We Create What We See* (New York: W. W. Norton & Co., 1998). A description and analysis of how the eye and brain actively construct a mental version of the external world.

Kandel, Eric R., *In Search of Memory: The Emergence of a New Science of Mind* (New York: W. W. Norton & Co., 2006). Part autobiography by a Nobel-winning neuropsychiatrist, part explanation of the brain mechanics and importance of memories.

LeDoux, Joseph, *The Emotional Brain: The Mysterious Underpinnings of Emotional Life* (New York: Simon & Schuster Paperbacks, 1996). An examination of emotion and its key role in making good decisions as well as basic survival.

Lehrer, Jonah, *How We Decide* (Boston: Mariner Books, 2009). A summary of current neuroscientific research about how the brain reaches decisions, augmented by case studies of good decision-making under stress.

Loftus, Elizabeth, and Katherine Ketcham, *Witness for the Defense: The Accused, the Eyewitness, and the Expert Who Puts Memory on Trial* (New York: St. Martin's Press, 1991). A personal and anecdotal account of the foibles of eyewitness testimony by a noted memory expert.

Ninio, Jacques, *The Science of Illusions* (Ithaca, NY: Cornell University Press, 2001). Translation of the original French edition. A well-illustrated spectrum of optical illusions, categorized and explained.

Purves, Dale, and R. Beau Lotto, *Why We See What We Do: An Empirical Theory of Vision* (Sunderland, MA: Sinauer Associates, 2003). An examination of a broad range of evidence suggesting the mind uses a statistical model, based on experience, to assign meaning to visual stimuli. Includes illusions of color and brightness.

Schacter, Daniel L., *The Seven Sins of Memory: How the Mind Forgets and Remembers* (Boston: Houghton Mifflin, 2001). Common problems, explained with concrete examples, that offer insight into the complex brain functions related to memory.

Schwartz, Barry, *The Paradox of Choice: Why More Is Less* (New York: Harper Perennial, 2004). A primer on good decision-making, with scientific explanation of the causes of mental stress from an overabundance of choices.

Shepard, Roger N., *Mind Sights* (New York: W. H. Freeman and Company, 1990). Optical illusions, including many whimsical ones, with commentary and explanation by their creator.

Wegner, Daniel W., *The Illusion of Conscious Will* (Cambridge, MA: Bradford Books, 2002). An exploration of how humans experience their own will and how the brain creates the feeling of being in control of one's actions.

Wilson, Timothy D., *Strangers to Ourselves: Discovering the Adaptive Unconscious* (Cambridge, MA: Belknap Press, 2002). Explores the role of the unconscious mind in making decisions, forming opinions and feelings, and solving problems.

Illustrations credits

1, PASIEKA/SPL/Getty Images; 2-3, Mark Grenier/Shutterstock; 4, Classic Images/Alamy; 6, courtesy David Copperfield; 8, Library of Congress; 11, CBS Photo Archive/Getty Images; 12, Mopic/Shutterstock; 14, Don Bliss/NIH Medical Arts; 15, Gary Carlson/Photo Researchers, Inc.; 16, Don W. Fawcett/Photo Researchers, Inc.; 17, Science Picture Co./Science Faction; 18, Science Picture Co./Science Faction; 19 (UP), Guigoz/Steiner/Photo Researchers, Inc.; 19 (CTR), Guigoz/Steiner/Photo Researchers, Inc.; 19 (LO), Guigoz/Steiner/Photo Researchers, Inc.; 20 (UP LE), Richard Tibbits/Antbits Illustration; 20 (UP RT), Richard Tibbits/Antbits Illustration; 20 (LO LE), Richard Tibbits/Antbits Illustration; 20 (LO RT), Richard Tibbits/Antbits Illustration; 22-23, S. Ragets/Shutterstock; 24, Tatiana Makotra/Shutterstock; 26, majeczka/Shutterstock; 27, Jireh Design; 28, Omikron/Photo Researchers, Inc.; 29, BSIP/Photo Researchers, Inc.; 31, Dorling Kindersley; 32, Paul Fleet/Shutterstock; 33, George Eade; 34, Paul Fleet/Shutterstock; 35, ©1999 Terese Winslow; 36, Noelle Weber; 38, Noelle Weber; 39 (LE), Rich Reid/NationalGeographicStock.com; 39 (UP RT), gaspr13/iStockphoto.com; 39 (RT CTR), Tim Davis/CORBIS; 39 (LO RT), Allison Michael Orenstein/Getty Images; 40, public domain; 41 (UP), Robert Spriggs/Shutterstock; 41 (LO), Noelle Weber; 42, Adam Radosavljevic/Shutterstock; 43 (LE), John Wollwerth/Shutterstock; 43 (RT), Kamira/Shutterstock; 44, Noelle Weber; 46 (LE), Noelle Weber; 46 (RT), Noelle Weber; 48, Raychel Deppe/iStockphoto.com; 49, manzrussali/Shutterstock; 50 (UP), Created by John Sadowski based on original photo by Tomas Sereda/Shutterstock; 50 (LO), Created by John Sadowski based on original photo by Tomas Sereda/Shutterstock; 52-53, James P. Blair; 54, William Whitehurst/CORBIS; 55 (LE), Ted Kinsman/Photo Researchers, Inc.; 55 (RT), Ted Kinsman/Photo Researchers, Inc.; 56, PZDesigns/Shutterstock; 57, Noelle Weber; 59, Noelle Weber; 60, Image by R. Beau Lotto of www.lottolab.org; 62, Image by R. Beau Lotto of www.lottolab.org; 64, Noelle Weber; 65, alfredolon/Shutterstock; 66, Camilo Vergara/Photo Researchers, Inc.; 67, Noelle Weber; 68, Noelle Weber; 70, Noelle Weber; 71, North Coast Medical, Inc.; 72, Joachim Wendler/Shutterstock; 73, Dr. Paul V. Maximov, Laboratory of Sensory Information Processing, Institute for Information Transmission Problems, Russian Academy of Sciences; 74, The Forest Has Eyes, ©Bev Doolittle, courtesy of The Greenwich Workshop®, Inc.; 76, The Forest Has Eyes, ©Bev Doolittle, courtesy of The Greenwich Workshop®, Inc.; 77, XAOC/Shutterstock; 78 (UP), Helder Almeida/Shutterstock; 78 (LO), Big Pants Production/Shutterstock; 78-79 (UP), Matt Ragen/Shutterstock; 78-79 (LO), Kevin H. Knuth/Shutterstock; 79 (UP), Yvan/Shutterstock; 79 (LO), JR Trice/Shutterstock; 80 (LE), Chepko Danil Vitalevich/Shutterstock; 80 (RT), Khoroshunova Olga/Shutterstock; 81 (UP), The Forest Has Eyes, ©Bev Doolittle, courtesy of The Greenwich Workshop®, Inc.; 81 (LO), The Forest Has Eyes, ©Bev Doolittle, courtesy of The Greenwich Workshop®, Inc.; 82, ©1995, Edward H Adelson; 84, ©1995, Edward H Adelson; 85, ULTRA.F/Getty Images; 86, cobalt88/Shutterstock; 87, David Mack/Photo Researchers, Inc.; 88, Erich Lessing/Art Resource, NY; 90, Erich Lessing/Art Resource, NY; 91, Wayne Johnson/Shutterstock; 92, Hannah Eckman/Shutterstock; 93, Images.com/CORBIS; 94, Close, Chuck (1940-) © Copyright. Self-Portrait. 1997. Oil on canvas, 8' 6" x 7' (259.1 x 213.4 cm). Gift of Agnes Gund, Jo Carole and Ronald S. Lauder, Donald L. Bryant, Jr., Leon Black, Michael and Judy Ovitz, Anna Marie and Robert F. Shapiro, Leila and Melville Straus, Doris and Donald Fisher, and purchase. The Museum of Modern Art, New York, NY, U.S.A. Digital Image © The Museum of Modern Art/Licensed by SCALA/Art Resource, NY; 98, Brian A. Jackson/Shutterstock; 100, Juan Velasco; 101, Laughing Stock/CORBIS; 102, caimacanulShutterstock; 103 (UP), Juan Velasco; 103 (LO), Juan Velasco; 104, Cary Sol Wolinsky; 105, 2happy/Shutterstock; 106, Ira Block; 107, ©Sam Gross/The New Yorker Collection/www.cartoonbank.com; 108 (UP), Evgeny Murtola/Shutterstock; 108 (LO), Evgeny Murtola/Shutterstock; 109, Smit/Shutterstock; 111, Time & Life Pictures/Getty Images; 112, Athanasia Nomikou/Shutterstock; 113, Ugorenkov Aleksandr/Shutterstock; 114, Ugorenkov Aleksandr/Shutterstock; 120, David Svetlik/Shutterstock; 121, Pgiam/iStock, Noelle Weber and Susan Blair; 125, Images.com/CORBIS; 126, Cary Sol Wolinsky; 127, EpicStockMedia/Shutterstock; 128, Anton Prado PHOTO/Shutterstock; 130, silver-john/Shutterstock; 131, Ikon Images/CORBIS; 132, Eric Isselée/Shutterstock; 133, Elena Elisseeva/Shutterstock; 134 (LO LE), IKO/Shutterstock; 134 (UP RT), sepavo/Shutterstock; 135, Morgan Lane Photography/Shutterstock; 136-137, Natalie-Claude zbelanger/iStockphoto.com; 138, Fred Bavendam/Minden Pictures/National GeographicStock.com; 140-141, Strejman/Shutterstock; 144, Dudarev Mikhail/Shutterstock; 145, Pinon Road/Shutterstock; 146, s duffett/Shutterstock; 148, Anton Seleznev/iStockphoto.com; 149, Elena Moiseeva/Shutterstock; 150, Bliznetsov/Shutterstock; 151 (LE), S1001/Shutterstock; 151 (RT), S1001/Shutterstock; 152 (LE), Portrait of the Young Voltaire (1694-1778) (oil on canvas), French School, (18th century) / Musee Antoine Lecuyer, Saint-Quentin, France / Giraudon / The Bridgeman Art Library International; 152 (RT), Portrait of Henry VIII, c.1540 (oil on panel), Holbein the Younger, Hans (1497/8-1543) (circle of) / Private Collection / Photo © Philip Mould Ltd, London / The Bridgeman Art Library International; 153, 3d kot/Shutterstock; 156, Noelle Weber; 157, Noelle Weber; 158, Roman Sigaev/Shutterstock; 159, Lucs-kho/Wikipedia; 160, Undergroundarts.co.uk/Shutterstock; 162-163, Feng Yu/Shutterstock; 164, Dvirus/Shutterstock; 165, MediVisuals/Photo Researchers, Inc.; 166, Laurent Renault/Shutterstock; 167, Images.com/CORBIS; 168, Carlyn Iverson/Photo Researchers, Inc.; 169, Rebecca Hale, NGS; 171, Conde Nast Archive/CORBIS; 172, Wolfgang Schaller/Shutterstock; 173, Undergroundarts.co.uk/Shutterstock; 174, Pixelbliss/Shutterstock; 175, Audrey Snider-Bell/Shutterstock; 176, Robert F. Balazik/Shutterstock; 177, Iurii Osadchi/Shutterstock; 178, iconspro/Shutterstock; 180, yamix/Shutterstock; 181, Aspect3D/Shutterstock; 182, nutech21/Shutterstock; 183, Betacam-SP/Shutterstock; 184, Library of Congress; 186, Olegro/Shutterstock; 188, Maggie Steber; 189, Stephen Mcsweeny/Shutterstock; 190, El Greco/Shutterstock; 191, Images.com/CORBIS; 192, Mark Thiessen, NGP; 194, higyou/Shutterstock; 195, Sergei Karpukhin/Reuters/CORBIS; 196, Ken Shimizu; 197, AP Images/Seth Wenig; 198, MGM/The Kobal Collection; 199, Oliver Widder; 200, Phase4Photography/Shutterstock; 202, Paul Bodea/Shutterstock; 204, K.Jakuowska/Shutterstock; 205 (LE), Juanmonino/iStockphoto.com; 205 (RT), Juanmonino/iStockphoto.com; 206, Anton Balazh/Shutterstock; 208, Gregor Kervina/Shutterstock; 211, Ocean/CORBIS; 213, AP Images/Courtesy of Harvard Medical School; 214-215, Images.com/CORBIS; 216, Fotocrisis/Shutterstock; 219, SVLuma/Shutterstock.

About the Consultant

Cedar Riener received his A.B. in the history of science from Harvard College and his Ph.D. in cognitive psychology from the University of Virginia. He is currently an assistant professor of psychology at Randolph-Macon College in Ashland, Virginia, where he has designed a course called "The Psychology of Illusions," from which many of this book's examples are drawn.

Brainworks

Michael S. Sweeney

The National Geographic Society is one of the world's largest nonprofit scientific and educational organizations. Founded in 1888 to "increase and diffuse geographic knowledge," the Society's mission is to inspire people to care about the planet. It reaches more than 400 million people worldwide each month through its official journal, *National Geographic,* and other magazines; National Geographic Channel; television documentaries; music; radio; films; books; DVDs; maps; exhibitions; live events; school publishing programs; interactive media; and merchandise. National Geographic has funded more than 9,600 scientific research, conservation and exploration projects and supports an education program promoting geographic literacy.

For more information, please call 1-800-NGS LINE (647-5463) or write to the following address:

National Geographic Society
1145 17th Street N.W.
Washington, D.C. 20036-4688 U.S.A.

Visit us online at www.nationalgeographic.com

For information about special discounts for bulk purchases, please contact National Geographic Books Special Sales: ngspecsales@ngs.org

For rights or permissions inquiries, please contact National Geographic Books Subsidiary Rights: ngbookrights@ngs.org

Library of Congress Cataloging-in-Publication Data
Sweeney, Michael S.
Brainworks: The mind-bending science of how you see, what you think, and who you are / by Michael S. Sweeney; foreword by illusionist David Copperfield.
p. cm.

ISBN 978-1-4262-0757-0 (hardcover: alk. paper)

1. Thought and thinking. 2. Self.
3. Brain. I. Title.
BF441.S94 2011
153--dc22
2011011780

Printed in the United States of America
11/QGT-CML/1